Organic Chemistry Concepts and Applications for Medicinal Chemistry

Organic Chemistry Concepts and Applications for Medicinal Chemistry

JOSEPH E. RICE
Associate Professor of Medicinal Chemistry
Department of Medicinal Chemistry
Ernest Mario School of Pharmacy
Rutgers, The State University of New Jersey
Piscataway, NJ 08854-8020

AMSTERDAM • BOSTON • HEIDELBERG • LONDON
NEW YORK • OXFORD • PARIS • SAN DIEGO
SAN FRANCISCO • SINGAPORE • SYDNEY • TOKYO
Academic Press is an imprint of Elsevier

Academic Press is an imprint of Elsevier
525 B Street, Suite 1900, San Diego, CA 92101-4495, USA
225 Wyman Street, Waltham, MA 02451, USA

Notice
No responsibility is assumed by the publisher for any injury and/or damage to persons or property as a matter of products liability, negligence or otherwise, or from any use or operation of any methods, products, instructions or ideas contained in the material herein. Because of rapid advances in the medical sciences, in particular, independent verification of diagnoses and drug dosages should be made.

Library of Congress Cataloging-in-Publication Data
Rice, Joseph E., author.
   Organic chemistry concepts and applications for medicinal chemistry / Joseph E. Rice, Rutgers, the State University of New Jersey.
      pages cm
   Includes bibliographical references (pages).
   ISBN 978-0-12-800739-6
1. Clinical chemistry. 2. Biochemistry. 3. Organic compounds. I. Title.
   QD415.R48 2014
   547--dc23
                           2014005300

British Library Cataloguing in Publication Data
A catalogue record for this book is available from the British Library

ISBN: 978-0-12-800739-6

For information on all Academic Press publications
visit our web site at store.elsevier.com

Printed and bound in USA
14 15 16 17 18   10 9 8 7 6 5 4 3 2 1

Working together
to grow libraries in
developing countries

www.elsevier.com • www.bookaid.org

DEDICATION

This book is dedicated to my wife Dorothy and my daughters Jennifer and Christina.

CONTENTS

Medicinal chemistry is a discipline that is focused on the relationship between chemical structure and biological activity. This relationship serves as the basis for the design and synthesis of new drug entities. Knowledge of medicinal chemistry is therefore of utmost importance to students in pharmacy programs, to medical and nursing students, and to chemists involved in the drug design process.

Pharmacy students at Rutgers take organic chemistry as sophomores but are not taught medicinal chemistry until their fourth year. Shortly after I started teaching medicinal chemistry I noticed that many students had difficulty recalling some of the important concepts from two years prior. To address that need I wrote an extended handout to serve as a review. Over the years that handout has been revised and amended numerous times. This book grew out of those efforts.

The first four chapters provide a review of important concepts first introduced in organic chemistry. Among the topics covered are chemical bonding, stereochemistry and conformation, functional groups and acid–base chemistry. It is essential that the students thoroughly understand these principles so as to be able to fully grasp the subtle relationship between chemical structure and pharmaceutical activity that is at the core of medicinal chemistry. Reaction mechanisms and chemical transformations, while certainly of great importance to those involved in the synthesis of new drug entities, are not discussed in this book.

Chapter five introduces the student to partition coefficients, a subject not likely covered in undergraduate organic chemistry courses. The ease with which a drug passes from one compartment of the body to another is related to its partition coefficient. Like every other property, partition coefficients depend on structure with each portion of a molecule contributing to its overall hydrophobicity or hydrophilicity.

An integral part of medicinal chemistry education at Rutgers is training students to draw the correct structure of a drug when presented with its systematic name. The sixth chapter of this book is therefore concerned with nomenclature. While the naming of simple compounds such as alkanes, alkenes, alkynes, alcohols, amines, and carbonyl derivatives is introduced in almost all organic chemistry courses, the systematic nomenclature of heterocyclic compounds and complex polycyclic ring systems is generally not covered. Chapter six begins with a discussion of relatively simple heterocyclic ring systems and then progresses to more complex compounds that involve fused, bridged, and spiro polycyclic compounds. Such ring systems form the basis of a majority of drug structures. A series of rules for numbering these ring systems is described. Methods are also described for specifying the attachment of highly complex substituents onto parent compounds. The last section of this chapter focuses on the special nomenclature and numbering of a few natural product classes of medicinal interest including steroids, prostaglandins, and morphinans.

The chemistry of drug metabolism is reviewed in the final chapter of this book. Most foreign chemicals become transformed once they enter the body. In this process, which is known as metabolism, some existing functional groups on compounds undergo chemical changes while in other instances new functionality is added. Most commonly this is done to aid in the removal of the foreign substance from the body. The reader will learn to associate a variety of structural features of drug molecules with likely metabolic pathways and will become familiar with the chemical processes that are involved in the formation of specific metabolites.

Joseph E. Rice
Ernest Mario School of Pharmacy
Rutgers, The State University of New Jersey

CHAPTER 1

Bonding in Organic Compounds

By the time most students in the pharmaceutical sciences get to their third or fourth year, they have very likely already taken general chemistry and organic chemistry and have been exposed to the concepts of chemical bonding several times. The reader is likely asking: Why waste time going over this again when there are more interesting topics to explore? After all, it's pretty simple. Electrons are shared between two atoms to form bonds. There are single bonds, double bonds, and triple bonds. Can we now move on and get to something new?

The truth is that one cannot begin to understand organic chemistry unless he/she has a thorough understanding of the bonds that hold organic compounds together. It is the bonds that determine the three-dimensional shapes of molecules. With many drug molecules, shape is a factor that determines how the drug molecule interacts with its receptor. The bonds in a molecule also determine its chemical reactivity and properties. Again with drug molecules, reactivity relates to its chemical stability and to the types of metabolic transformations the drug may undergo. Bonds affect the acid/base properties of a compound and also affect the solubility. For drugs, the relative solubility in water and lipids affects how the molecule passes from one bodily compartment to another.

ATOMIC ORBITALS

Most drugs are organic compounds and have structures that are largely composed of carbon atoms. Carbon (atomic number 6) has a nucleus containing six positively charged protons. As a counterbalance, six negatively charged electrons orbit the nucleus. Electrons are restricted to *orbitals*, which are the functions that represent the probability of finding a given electron in space. Any orbital can be occupied by a maximum of two electrons, and these must have opposite spin quantum numbers (represented as ↑ or ↓). The shapes of orbitals are designated with small letters such as s, p, d, or f. Elements that are in the first row of the Periodic Table (hydrogen and helium) have only a single shell of electrons that reside in an s-orbital. In the second row, two shells exist. The second shell being larger than the first can contain more electrons, which are held in one s-orbital and three p-orbitals. Third row elements have three shells of electrons with the third now capable of having five d-orbitals in addition to

Organic Chemistry Concepts and Applications for Medicinal Chemistry
http://dx.doi.org/10.1016/B978-0-12-800739-6.00001-2

Table 1.1 The First Three Rows of the Periodic Table

	I	II		III	IV	V	VI	VII	VIII
1	H								He
2	Li	Be		B	C	N	O	F	Ne
3	Na	Mg		Al	Si	P	S	Cl	Ar

the one s- and three p-orbitals. Thus the row number in the Periodic Table tells how many shells of electrons a particular element contains (Table 1.1).

Electrons in s-orbitals are distributed spherically symmetrical about the nucleus (like a hollow ball with the nucleus at the center). Carbon, being found in the second row of the Periodic Table, has two shells of electrons. Two electrons are contained within the 1s orbital and the remaining four are found in the orbitals of the second shell. Note the fact that carbon is located in Group IV of the Periodic Table and has four electrons in the outer shell or *valence shell*. Electrons completely populate lower energy orbitals before higher energy orbitals. This is known as the *Aufbau principle*. Thus the 1s orbital, being closer to the nucleus, is filled before the 2s orbital. Within a given shell, the s-orbital is lower in energy than the p-orbitals. The three p-orbitals are dumbbell shaped and are mutually perpendicular to one another, directed along the x-, y-, and z-axes (p_x, p_y, and p_z). They are also equivalent in energy (*degenerate*). The remaining valence electrons first fill the 2s orbital and then partially fill two of the degenerate p-orbitals. *Hund's Rule* states that electrons must be placed singly into degenerate orbitals before they can be paired. Thus the *electronic configuration* for carbon becomes $1s^2 2s^2 2p_x^1 2p_y^1$. In general, only electrons in the valence shell participate in chemical bonding (Figure 1.1).

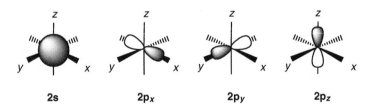

| 2s | 2p$_x$ | 2p$_y$ | 2p$_z$ |

Figure 1.1 Valence shell orbitals of second row elements.

Elements in the second row of the Periodic Table have four orbitals in the valence shell and these can hold up to a total of eight electrons (Table 1.2). Neon, with each of its orbitals completely filled, cannot accept any more electrons and is reluctant to give any away. It is therefore chemically inert, as are other members of Group VIII. As you move from left to right

Table 1.2 Electronic Configurations for the Second-Row Elements

Li	$2s^1$	Group I
Be	$2s^2$	Group II
B	$2s^2 2p_x^1$	Group III
C	$2s^2 2p_x^1 2p_y^1$	Group IV
N	$2s^2 2p_x^1 2p_y^1 2p_z^1$	Group V
O	$2s^2 2p_x^2 2p_y^1 2p_z^1$	Group VI
F	$2s^2 2p_x^2 2p_y^2 2p_z^1$	Group VII
Ne	$2s^2 2p_x^2 2p_y^2 2p_z^2$	Group VIII

across the Periodic Table within a particular row, each nucleus has one more proton than its neighbor at the left. The nuclear charge increases accordingly, but the distance from the nucleus to the valence electrons remains about the same. Thus the attraction of the nucleus for the valence electrons increases from left to right. This is known as *electronegativity*. Electronegativity also increases as you move up the Periodic Table within any given group (column) [as you move up in any group, the valence shell electrons are getting closer to the nucleus and are thus pulled at by the nucleus to a greater extent]. Fluorine is the most electronegative element (4 on a scale of 0–4).

Now that we know something about carbon it is time to start combining it with other elements to form molecules. The simplest organic compound is methane, CH_4, so let us see how it is constructed. Organic compounds are held together primarily by *covalent bonds*. Such bonds are formed whenever two valence electrons are shared between adjacent atoms. Since the $2p_x$ orbital of carbon has a single electron, it can overlap with a hydrogen 1s orbital, which also has one electron, and they can form a C–H bond. A similar scenario can be achieved with the $2p_y$ orbital of carbon and another hydrogen atom. However, a problem now arises when trying to bond the remaining two hydrogens with carbon. The $2p_z$ orbital is vacant, and overlapping with a 1s orbital of hydrogen would not result in a bond being formed because a covalent bond requires *two* electrons to be shared. Likewise, the carbon 2s orbital is completely filled, and overlapping with a hydrogen 1s orbital would result in two atoms sharing three electrons, which is also impossible. Thus if we use only the atomic orbitals of carbon to bond with hydrogen, we can form only CH_2 and predict that it would have a H-C-H bond angle of 90°. As anybody who has ever driven past a landfill can attest to, however, methane is a very real compound. Moreover, it has been established that CH_4 is tetrahedral in shape. The question is, therefore, how can we account for both the composition of methane and its shape (Figure 1.2)?

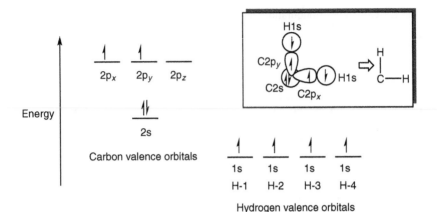

Figure 1.2 The atomic orbitals involved in forming methane. Inset box shows a pictorial representation of the results of using the carbon atomic orbitals to bond with hydrogen.

HYBRID ORBITALS

Orbitals, as mentioned previously, are probability functions. Linus Pauling recognized that as functions they can be manipulated mathematically by taking linear combinations [1]. When combining four functions, one obtains four new functions. In the case of the valence orbitals of carbon, those four functions are the 2s, $2p_x$, $2p_y$, and $2p_z$ orbitals. These are combined as follows:

$$2s + 2p_x + 2p_y + 2p_z = sp^3 \text{ hybrid orbital}$$

$$2s + 2p_x + 2p_y - 2p_z = sp^3 \text{ hybrid orbital}$$

$$2s + 2p_x - 2p_y + 2p_z = sp^3 \text{ hybrid orbital}$$

$$2s - 2p_x + 2p_y + 2p_z = sp^3 \text{ hybrid orbital}$$

Each of the resulting hybrid orbitals have 25% s–character and 75% p–character and so look like distorted dumbbells, with one lobe substantially larger than the other. In addition, the four new sp^3 hybrid orbitals are directed to the corners of a tetrahedron (Figure 1.3).

The newly formed sp^3 hybrid orbitals are degenerate and hence each is populated by a single electron from the carbon's four valence electrons (Figure 1.4). Since there are now four singly populated orbitals, it becomes easy to see how they can each overlap with a hydrogen 1s orbital to form four C–H bonds. Also, since those orbitals point to the corners of a tetrahedron, the three-dimensional shape of methane must also be tetrahedral. The

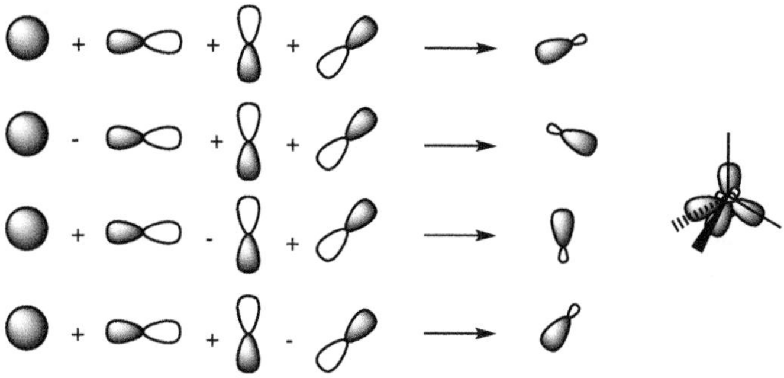

Figure 1.3 Pictorial representation of taking a linear combination of the four carbon valence orbitals to form sp³-hybrid orbitals.

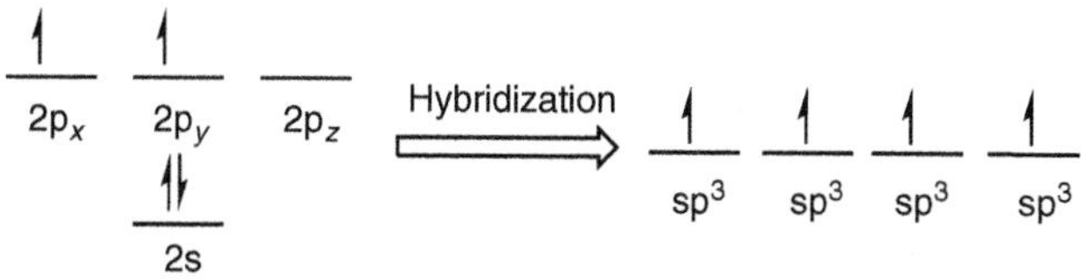

Figure 1.4 Result of hybridization of all four of the carbon valence orbitals.

geometry places the four hydrogen atoms as far away from each other as possible while still maintaining a bonding relationship with the central carbon. This minimizes electron repulsions from one bond to another and is known as the *Valence-Shell Electron-Pair Repulsion (VESPR)* rule. When four equivalent bonds are formed with carbon, the bond angle approaches the ideal 109.5° angle of a regular tetrahedron and the bond lengths will be identical (Figure 1.5). If the groups are different, then there will be deviations from this angle, as well as differences in their bond lengths. It will be seen later that there are often large deviations from this ideal bond angle in three- and four-membered rings. These deviations have a pronounced affect on the chemical reactivity of such compounds.

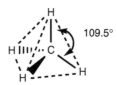

Figure 1.5 The tetrahedral structure of methane.

The overlap of a hybrid orbital of carbon with a hydrogen s-orbital forms what is known as a σ-*bond* (sigma bond). If one were to connect the two nuclei involved in a σ-bond with an imaginary line, the electron density would be distributed cylindrically symmetrical about that line. In a similar manner, σ-bonds can also be formed if the two hybrid orbitals overlap head-to-head, or if two s-orbitals overlap (Figure 1.6). Sigma bonds are usually depicted as a single line joining two atom symbols (C–H). While this is convenient, it must be remembered that the line drawn represents a pair of electrons. It is not a rigid spacer that connects the atoms, but something that can be bent, compressed, and stretched depending on the forces that it encounters. It is also not necessary for the electrons in a bond to be equally shared by both nuclei. If one atom is more electronegative than the other, the electrons will lie closer to the more electronegative atom and the bond will be *polarized*. The end with greater electron density will have a partial negative charge, whereas the opposite end will be electron deficient and has a partial positive charge. The distribution of electron density within bonds determines many of the chemical and physical properties of organic compounds.

Just as the four valence atomic orbitals of carbon were mixed to form a new set of four hybrid orbitals centered on carbon, the process of covalent bond formation also involves orbital mixing. To form a C–H bond, a hybrid orbital of carbon is mixed with the 1s orbital of hydrogen. The result is a new orbital called a *σ-orbital*, and it is lower in energy than either the hybrid orbital or the 1s orbital. One electron in the sp^3 hybrid orbital and the one in the 1s orbital can both be placed into the new σ-orbital, which is called a *bonding molecular orbital (MO)*. However, because two orbitals were combined, two new orbitals must be formed. Hence in addition to the bonding σ-orbital, a so-called *antibonding σ*-molecular orbital* is also formed. The σ*-orbital is higher in energy than either the sp^3 or s-orbital, and is not populated with any electrons under ordinary conditions. When electrons occupy only bonding MOs, the resulting system is stabilized. Electrons placed into antibonding MOs destabilize a bond by increasing its energy. Therefore,

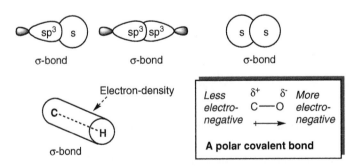

Figure 1.6 Various types of σ-bonds.

stable bonds will be formed when the total number of electrons in the overlapping atomic or hybrid orbitals is two. This can be realized by the overlap of two orbitals, each occupied by a single electron or by one orbital having two electrons overlapping with a vacant orbital (note that orbitals exist even if they have no electrons). An orbital with two electrons will not form a stable bond with an occupied orbital. This is because after filling the bonding molecular orbital, the remaining electrons would have to be placed into antibonding orbitals (Figures 1.7 and 1.8).

The molecule ethylene, C_2H_4, consists of an array of two carbon and four hydrogen atoms, in which each carbon is bonded to the other carbon and two hydrogens. Unlike methane, there are only three other atoms attached to carbon and therefore the geometry around each carbon is different than in methane. The VESPR rule requires that the three attached atoms be arranged in the same plane as the carbon and spaced approximately 120° apart from one another to minimize repulsions between the bonding electrons. This cannot be achieved using sp^3 hybrid orbitals and so a different hybridization scheme is required, one that will result in three

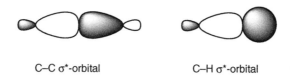

C–C σ*-orbital   C–H σ*-orbital

Figure 1.7 Pictorial representation of antibonding orbitals. Note that the orbitals interact out-of-phase.

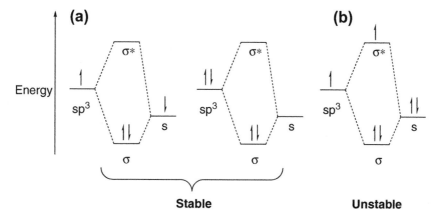

Figure 1.8 (a) Overlap of two partially filled or one filled and one vacant orbital to form a stable σ-bond. (b) Overlap of a filled and a partially filled orbital to form an unstable system.

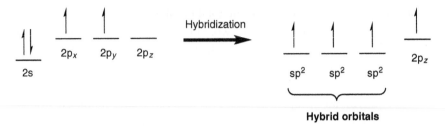

Figure 1.9 Mixing of the carbon 2s, $2p_x$, and $2p_y$ orbitals to create three sp^2-hybrid orbitals.

hybrid orbitals, not four. Such a set of orbitals is obtained by mixing the 2s orbital with two of the three 2p orbitals to give a set of three sp^2 hybrid orbitals. Since any two p-orbitals will reside in a plane, the resulting hybrid orbitals will also be planar and directed to the vertices of an equilateral triangle. As three singly occupied hybrid orbitals will be needed to form the three covalent bonds from carbon, each is populated by a single valence shell electron. The remaining electron remains in the unused $2p_z$ atomic orbital which of necessity must be perpendicular to the plane of the three sp^2 hybrid orbitals (Figure 1.9).

Two sp^2 orbitals from each carbon now overlap with two hydrogen 1s orbitals to form C–H σ-bonds. The remaining sp^2 orbital overlaps head-to-head with an sp^2 orbital on the other carbon to form a C–C σ-bond. The p_z-orbital on each carbon now overlaps side-to-side to form a second bond between the carbons that is called a *π-bond* (a π-molecular orbital occupied by the two electrons from the p_z orbitals). An antibonding *π* orbital* (unoccupied) is also formed at the same time. The carbons are joined by two different bonds and ethylene is said to possess a *double bond* with the formula often written as $H_2C\!\!=\!\!CH_2$. It is important to realize that the two bonds are not identical; one is a σ-bond and the other a π-bond. In ethylene, the H–C–H bond angle and the H–C–C bond angles are different, with the former being 118.7°and the latter 120.7°. If the C–C bond length (1.53 Å) in ethane (C_2H_6), in which both carbons are sp^3 hybridized, is compared to that of the ethylene 1.34 Å, it is seen that one effect of the π-bond in ethylene is to shorten the C–C bond length. In general, double bonds are shorter than the corresponding single bonds (Figures 1.10 and 1.11).

The presence of a double bond in a molecule has a great effect on its properties. In σ-bonds, the electron density is localized along the axis connecting the two atoms and is relatively unexposed. Electron density in a π-bond however is more accessible, being located above and below the

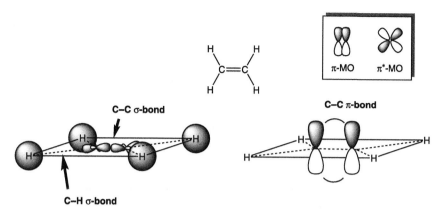

Figure 1.10 Representation of the two different types of bonds found in ethylene. Inset box shows a bonding and antibonding π-orbital.

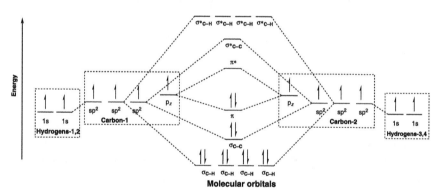

Figure 1.11 Molecular orbitals of ethylene formed as a result of bonding between two sp²-hybridized carbon atoms and four hydrogen atoms. Note that the π-orbital is higher in energy (more reactive) than the σ-orbitals.

plane of the molecule, and is therefore more susceptible to attack by *electrophiles* (electron–deficient reactive species). While atoms connected only by a σ–bond may rotate freely about the bond (like two wheels joined by an axle), this is not possible for atoms connected by double bonds. This allows for the existence of *geometrical isomers* (more about this later). If one examines a p-orbital (or a π-bond), it can be seen that there are two lobes of electron density, one above and the other below the plane of the molecule that decrease in size until they become zero in the plane of the molecule (a *node*). In practical terms, this means that there is a zero probability of finding an electron in the nucleus. One implication of this is that since electron density in p-orbitals decreases to zero in the plane where σ-bonds have

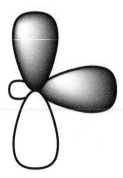

Figure 1.12 The p-orbital does not have any significant overlap with the hybrid orbital that is in a perpendicular plane.

their greatest electron density, there is no interaction (orbital overlap) between the electrons in p-orbitals (or π-bonds) and those in σ-orbitals. Put simply, orbitals that are perpendicular to one another do not overlap. Thus the π-electron system can be treated as being independent of the σ-electron system (Figure 1.12).

The molecule acetylene, C_2H_2, presents yet another bonding situation for carbon. In this case, each carbon is bonded to hydrogen and the other carbon. Acetylene and related compounds are linear, with the two groups attached to carbon 180° apart from each other. This arrangement cannot be obtained using sp^3 or sp^2 hybrid orbitals, and so when only two groups are attached to carbon, the atomic orbitals combine themselves in a different way to form a new set of hybrid orbitals. Mixing the 2s orbital with only one of the p-orbitals creates a set of two *sp hybrid orbitals* with an angle of 180°. For acetylene, overlap of one of the sp orbitals on each carbon with a hydrogen 1s orbital creates two new C–H σ-bonds. Overlap of the remaining sp orbital on each carbon produces a new C–C σ-bond. That only accounts for two of the valence electrons on each carbon. The remaining two are located, one each, in the remaining two p-orbitals. These p-orbitals are mutually perpendicular and are also perpendicular to the σ-bonding array (H–C–C–H). When the corresponding p-orbitals overlap side-to-side, they form two π-bonds. Therefore, the two carbons of acetylene are connected by three bonds—one σ-bond and two π-bonds. This is known as a *triple bond*, and is depicted as H–C≡C–H. The C–C bond length in acetylene is 1.18 Å which is shorter still than that of ethylene as a result of the additional π-bond. As with double bonds, triple bonds are also more reactive because the electron density is concentrated away from the interatomic bond axis (Figures 1.13 and 1.14).

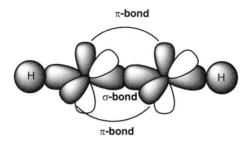

π-bond

σ-bond

π-bond

Figure 1.13 Bonding picture of acetylene showing the σ-bond and the two perpendicular π-bonds connecting the two carbon atoms.

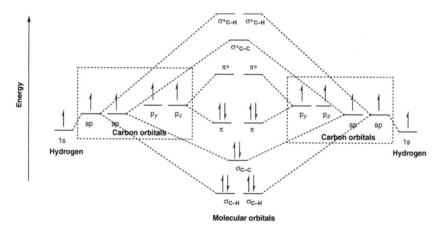

Figure 1.14 Molecular orbitals formed as a result of bonding between two sp-hybridized carbon atoms and two hydrogen atoms in acetylene. Note that the π-orbitals are higher in energy (more reactive) than the σ-orbitals.

The discussions on bonding to this point show that molecular geometry is dependent on hybridization. Whenever carbon is attached to four other groups, it organizes into sp³ hybrid orbitals and the geometry at that carbon is *tetrahedral*. This means that the carbon and any two of the other groups attached to it are all in the same plane. Of the remaining two groups attached to the carbon, one will be in front of the plane and the other will be behind the plane. It is common to try to depict such three-dimensional arrangements using either heavy solid lines or solid wedges for groups coming out from the paper and hashed lines or hashed wedges for those groups which are oriented behind the paper (Figure 1.15).

A carbon that is connected to only three other groups will be sp² hybridized and all three attached groups will be in one plane in a trigonal array

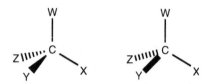

Figure 1.15 Tetrahedral geometry. In these drawings, the carbon and attached groups W and X are all in the plane of the paper. Group Y is in front of the plane (solid wedge or heavy line) and group Z is behind the plane (hashed wedge or hashed line).

(~120°) around the carbon. When only two groups are connected to the carbon, then the two groups and the carbon will be in a linear array with the bond angle between the two groups 180°. These relationships do not just pertain to carbon however. Other common elements can also form hybrid orbitals, with the same geometric relationships as carbon.

Nitrogen is next to carbon on the Periodic Table and can also reorganize its valence shell electrons into four sp^3 hybrid orbitals. As a Group V element, however, it has one more valence shell electron than carbon and therefore one of the sp^3 hybrid orbitals will be completely filled (Figure 1.16).

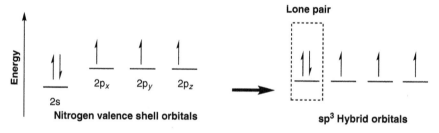

Figure 1.16 Reorganization of the valence shell atomic orbitals of nitrogen into four sp^3 hybrid orbitals.

The three partially filled orbitals can form three covalent bonds by overlapping with three partially filled s- or hybrid orbitals on other atoms. The remaining filled sp^3 orbital is known as a *lone pair*. Nitrogen can therefore form three covalent bonds to other atoms with a lone pair remaining associated with the nitrogen as the fourth group that completes the tetrahedral geometry (again the nitrogen and two of the attached groups are in one plane with one of the remaining groups in front of the plane and one behind the plane). The lone pair on nitrogen is responsible for the fact that many nitrogen compounds act as *bases* (more about this later). Since any chemical bond can only have two electrons, the only way the lone pair can form a bond is by overlapping with an unoccupied (vacant) orbital (for example,

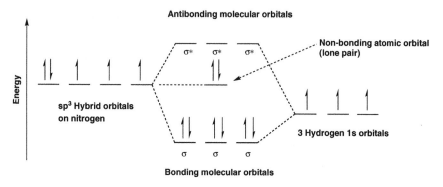

Figure 1.17 Orbital picture for ammonia (NH_3).

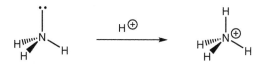

Figure 1.18 The reaction of ammonia with a proton (a hydrogen atom that has lost its electron and therefore has a vacant 1s orbital). Notice that the nitrogen becomes positively charged as a result of this reaction.

a proton which is a hydrogen atom that has lost its one electron). If this happens, the nitrogen will then acquire a *positive charge* because those electrons are no longer localized exclusively on the nitrogen rendering it electron deficient. The geometry at the nitrogen is still tetrahedral because the fourth group takes the place of the lone pair (Figures 1.17 and 1.18).

Oxygen has one more valence shell electron than nitrogen and therefore has two filled and two partially filled sp^3 hybrid orbitals. As a result, oxygen tends to form two covalent bonds with two lone pairs remaining on the oxygen atom. Because of the sp^3 hybridization, the geometry around oxygen is tetrahedral. The lone pairs of oxygen are not donated as readily to form bonds with vacant orbitals as with nitrogen because oxygen is much more electronegative than nitrogen (Figure 1.19).

Fluorine is in Group VII of the second-row elements and therefore has seven valence shell electrons. This gives rise to three filled sp^3 orbitals and one partially filled orbital. The result of this is that fluorine forms only one covalent bond and has three lone pairs of electrons arranged in a tetrahedral geometry. Due to its extremely high electronegativity, fluorine is very reluctant to donate its lone-pair electrons to vacant orbitals on other atoms (Figure 1.20).

A nitrogen atom that is attached to only three groups is sp^2 hybridized. One of the sp^2 orbitals is filled with two electrons (the lone pair) while the other two and the remaining p-orbital each have one electron. Upon

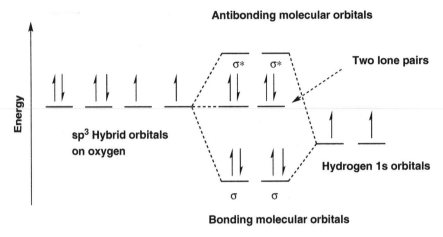

Figure 1.19 The orbital picture of water (H_2O).

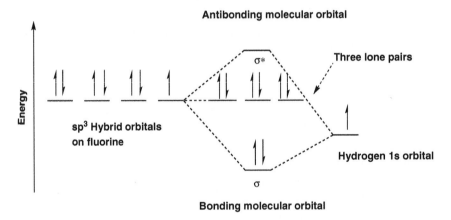

Figure 1.20 The orbital picture of hydrogen fluoride (HF).

forming a bond with another sp^2-hybridized atom (X), an N=X double bond is created (again, a σ-bond and a π-bond). The π-bond has little if any effect on the lone pair because they are in planes that are perpendicular, and therefore, the orbitals do not overlap (Figure 1.21).

If sp^2-hybridized nitrogen uses its lone pair to bond to another atom with a vacant orbital, the geometry at the nitrogen will still be planar, but the nitrogen will acquire a positive charge because those electrons are no longer associated exclusively with the nitrogen (Figure 1.22).

The situation in which oxygen is attached to only three groups is similar except that with the extra valence electron, two of the sp^2 hybrid orbitals contain lone pairs. The remaining sp^2 orbital can overlap with a similar

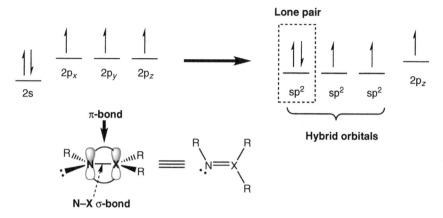

Figure 1.21 Hybrid orbitals (sp²) on nitrogen and the formation of N═X double bonds. Note that the lone pair and the π-bond are in perpendicular planes.

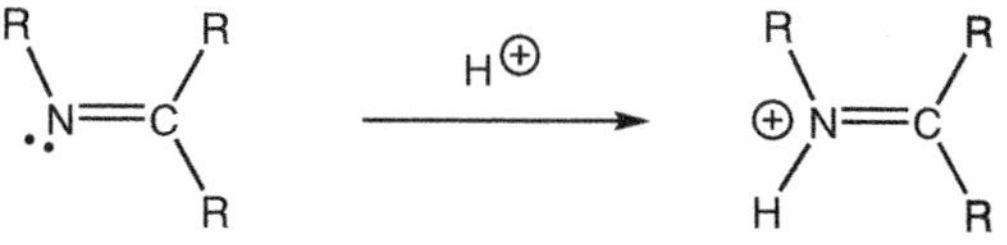

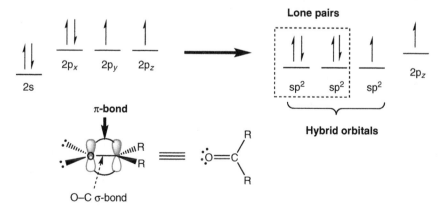

Figure 1.22 The reaction of an imine nitrogen with a proton to form an iminium cation.

orbital on another atom (carbon for example) to form a σ-bond. The adjacent p-orbitals can then overlap to form a π-bond giving rise to a C═O double bond. This bond is extremely common and is known as a *carbonyl bond* (Figure 1.23).

Figure 1.23 Oxygen sp² hybrid orbitals and the formation of carbonyl bonds. Both lone pairs are in a plane perpendicular to the π-bond.

Nitrogen can also form triple bonds with other atoms. In this case, the lone pair is located in an sp orbital on the nitrogen. A C≡N bond is known as a *nitrile* (or cyano group). Unlike sp²-hybridized nitrogen, the lone pair on the nitrogen of a nitrile is not readily donated to form bonds with atoms possessing vacant orbitals because the short bond length of the sp hybrid orbital keeps the lone pair close to the nitrogen (electronegativity in general increases as you go from sp³ to sp² to sp) (Figure 1.24).

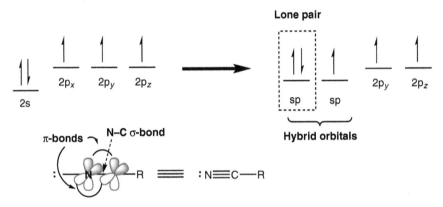

Figure 1.24 sp-Hybrid orbitals on nitrogen.

RESONANCE

There are no uncharged species with triple bonded oxygen. For oxygen to form a triple bond with carbon for example, one of the lone pairs must overlap with a vacant orbital on carbon (a cation). This leaves the oxygen electron deficient and it acquires a positive charge. The resulting species is called an *acylium ion*, and it is a highly reactive intermediate and not a stable organic compound. Since oxygen is more electronegative than carbon, a positive charge is better tolerated on carbon than on oxygen. Relocating a pair of electrons from one of the π-bonds back onto oxygen reestablishes a lone pair and creates a vacant orbital on carbon, imparting a positive charge. This is called an *acyl cation*. An acylium ion and an acyl cation are two species that have the same constitution but differ only in the way the electrons are distributed. These are referred to as *resonance structures*. Each of these structures has advantages and disadvantages: the acyl cation has the positive charge on the less electronegative element, which is good, but has a vacant orbital, which is bad [nature does not like a vacuum]; while the acylium ion has an extra bond, which is good, but has the positive charge on the more electronegative element, which is bad. Overall, the true picture

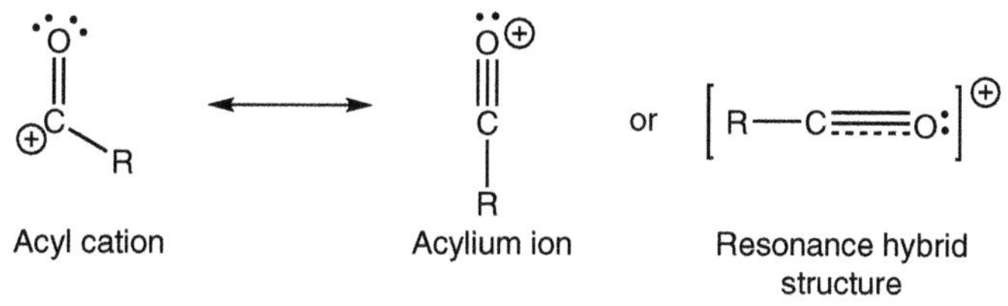

Figure 1.25 Acyl cation and acylium ion resonance structures.

of electron distribution in these species is a combination of these two extremes. It is a common practice to draw all possible resonance structures and join them with resonance arrows ($\leftrightarrow$). Alternatively, a composite resonance hybrid structure can be drawn (Figure 1.25).

When a charge is present in a compound, the ability of the molecule to delocalize that charge affects the total energy of the compound. In other words, a charge that is localized on a single atom raises the energy of a compound relative to a charge that can be spread out by resonance over several atoms. The example below will serve to illustrate this point.

Examination of the example in Figure 1.26 shows that for the top structure, the negatively charged oxygen is connected by a single bond to a

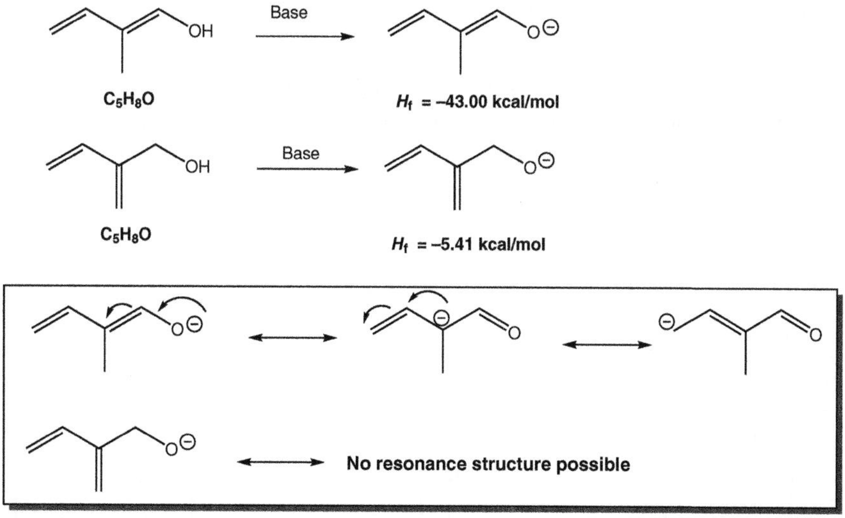

Figure 1.26 Treatment of isomeric alcohols with base generates two different alkoxide anions. The structure on top is 37.59 kcal/mol more stable (lower in energy) than the structure below. The reason for this great stability is that the negative charge in the top structure can be delocalized by resonance (see inset box), whereas the charge for the lower compound is localized on the oxygen. *Calculations of heats of formation (H_f) were performed using semiempirical molecular orbital methods (PM3), Spartan 06, Wavefunction, Inc.*

double bond that, in turn, is connected by another single bond and then to another double bond. This pattern of alternating single and double bonds is called a *conjugated system*. Conjugation acts much like a piece of copper wire does in conducting an electric current. It allows electrons to be conducted throughout the entire conjugated system. To see how this is possible, it is necessary to look at the orbitals involved in such a system. The oxygen is sp^3 hybridized and when a base removes a proton from the O–H bond, the electrons remain with the oxygen. Since the oxygen has two lone pairs plus one of the electrons from the C–O bond and one from the O–H bond, the addition of one extra electron from hydrogen gives a total of seven valence electrons (it should have six) and thus gives the oxygen a negative charge, represented as a filled sp^3 orbital. If the oxygen *rehybridizes* to sp^2, that extra electron pair will now reside in a p-orbital that can align with the rest of the p-orbitals of the carbons. The excess electrons can now be *delocalized* throughout the resulting conjugated system, landing on *every other* atom (O, C-2, and C-4) but not on C-1 and C-3. The energy gained by forming a conjugated system more than compensates for any energy expended in rehybridization of the oxygen. With the lower compound in Figure 1.26, however, the filled sp^3 orbital of the negatively charged oxygen has nowhere to go with its electrons because the carbon adjacent to it is sp^3 hybridized, and hence all of the orbitals are completely filled. In situations like this, the *saturated carbon* (sp^3 carbon) acts like an electrical insulator and prevents the conductance of electrons from oxygen to the carbon–carbon double bonds (Figure 1.27).

The compound benzene, C$_6$H$_6$, is cyclic and flat. Since each carbon is bonded to two carbons and one hydrogen, they are all sp^2 hybridized and each carbon also has a p-orbital that is partially filled with one electron. The flatness of the ring allows these orbitals to overlap to the maximum amount

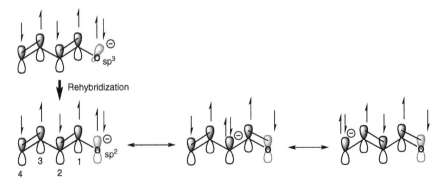

Figure 1.27 Orbital picture of the conjugated system from Figure 2.26.

to create three π-bonds separated from one another by σ-bonds—a conjugated system. The p-orbitals can also pair up in the opposite direction to give three other π-bonds. These two structures are in resonance and the actual distribution of electron density around the ring is a combination of these structures (Figure 1.28).

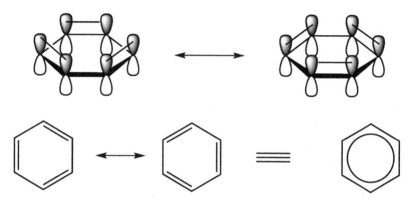

Figure 1.28 The major resonance structures of benzene (hydrogen atoms omitted for clarity). A circle inscribed in a hexagon is often used to represent all the resonance forms of benzene.

AROMATICITY

Benzene has three π-bonds, each with two electrons, for a total of six electrons in the π-system. Compounds such as benzene that are cyclic and flat, in which each atom in the ring contributes a p-orbital and that have six electrons in the π-system are said to be *aromatic*, and experience tremendous stabilization (energy lowering, ~36 kcal/mol) relative to the three isolated double bonds. The compound naphthalene is another example of aromaticity, this time with two fused aromatic rings. The entire structure is flat, each atom in each ring has a p-orbital with one electron, and therefore within each ring, the π-system contains six electrons. Note that the p-orbitals on the two carbons common to both rings (ring fusion positions) belong to both rings (Figure 1.29).

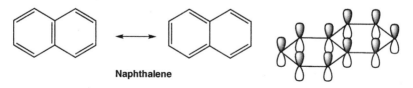

Naphthalene

Figure 1.29 The major resonance structures of naphthalene. Note that the two p-orbitals are shared by both rings.

Consider now the cycloheptatrienyl cation. This species has a seven-membered ring, with three conjugated double bonds. The remaining carbon is attached to only three other groups and so is sp² hybridized. The p-orbital however is vacant which places a positive charge on this carbon. Such a situation can arise if for example a group that is attached to a sp³-hybridized carbon leaves with both electrons from its former bond. In the cycloheptatrienyl cation, there are a total of six electrons in the π-system, each atom of the ring has a p-orbital, and the ring is flat. This therefore represents another aromatic ring system (Figure 1.30).

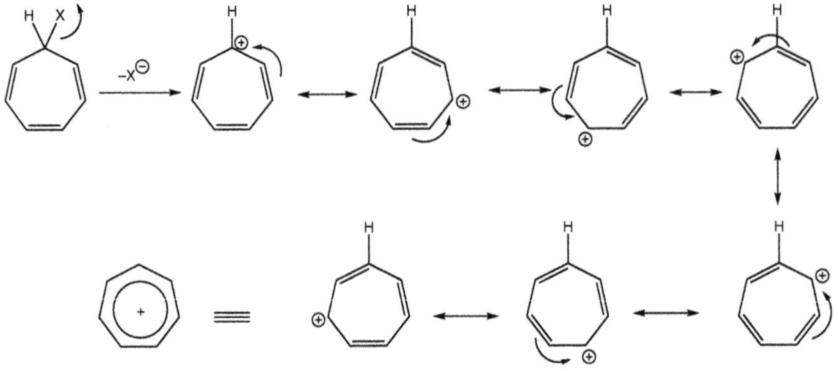

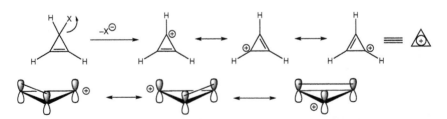

Figure 1.30 Resonance and aromaticity in the cycloheptatrienyl cation.

Another example of aromaticity is found in the cyclopropenium cation. This is a three-membered ring with one π-bond and a vacant p-orbital on the remaining carbon (creating a positive charge). The charge can then be distributed all around the ring by overlap of the vacant p-orbital with the p-orbitals from the π-bond. This ring system is flat and each atom in the ring has a p-orbital, but it only has two electrons in its π-system (Figure 1.31).

Figure 1.31 Resonance and aromaticity in the cyclopropenium cation.

The fact that six-electron systems and two-electron systems can be aromatic suggests that a more generalized definition of aromaticity be proposed. Thus for a system to be considered as aromatic, it must conform to each of the following requirements:

- It must by cyclic.
- It must be flat.
- Each atom in the ring must contribute a p-orbital.
- The total number of electrons in the π-system must be a solution to the expression $4n + 2$ where n is any integer $(0, 1, 2, 3...)$. Thus systems that satisfy the first three conditions and that have $2, 6, 10, 14...$ electrons in the π-system are aromatic. By far, the most common situation you will encounter is a system with six π-electrons.

In contrast to the compounds discussed above are those that fulfill the first three requirements for aromaticity, but have only four electrons in the π-system (actually $4n$ *systems* with $4, 8, 12...$ electrons). Such compounds do not experience any stabilization (energy lowering), but are actually *destabilized*. These are known as *antiaromatic* compounds. An example of an antiaromatic compound is cyclobutadiene, which can only be isolated and studied at extremely low temperatures because of its instability. An explanation for this phenomenon requires an understanding of orbital symmetry considerations, which is beyond the scope of this chapter (Figure 1.32).

At first glance, it would also appear that the compound cyclooctatetraene should also be antiaromatic since a two-dimensional representation appears to be a completely conjugated eight-electron system. This compound, however, is perfectly stable and readily available. The reason is that it is not flat, and thus resonance throughout the ring cannot occur. It behaves simply as a cyclic compound with four isolated double bonds and is neither aromatic nor antiaromatic (Figure 1.33).

Aromatic rings can also contain heteroatoms. Pyridine, C_5H_5N, for example, is a six-membered ring nitrogen *heterocycle* (contains an atom other than carbon in the ring). Like benzene, each atom of the ring is sp^2 hybridized, and

Cyclobutadiene
(Isolatable only at
extremely low temperatures)

-Flat
-Cyclic
-Each atom in the ring has a p-orbital
-Has 4 electrons in the π-system
Antiaromatic

Figure 1.32 Cyclobutadiene and antiaromaticity.

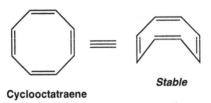

Cyclooctatraene   **Stable**

Figure 1.33 The structure of cyclooctatraene showing why it is not antiaromatic.

Pyridine

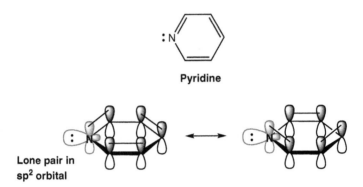

Lone pair in
sp² orbital

Figure 1.34 Pyridine as an aromatic heterocycle.

each therefore has a p-orbital with one electron that overlap to form three π-bonds. The ring is flat and has a total of six electrons in the π-system, and is aromatic. One might ask "What about the lone pair on nitrogen?" If the lone pair electrons were included together with the electrons already in the π-system, it would give a total of eight π-electrons, which would result in pyridine being an antiaromatic compound. The lone pair, however, cannot be included with the electrons in the π-system. Remember that for sp²-hybridized nitrogen, the lone pair is contained within a sp² hybrid orbital. Applied to pyridine, this means that the lone pair is in an orbital that is in the plane of the ring. The p-orbitals are perpendicular to this plane and cannot overlap with the sp² hybrid orbital. The lone pair can therefore act independent of the π-system, a fact that accounts for the propensity of that electron pair to react readily with most electrophiles, including protons (Figure 1.34).

Other six-membered ring nitrogen heterocycles that are fully unsaturated are also aromatic. Included here are compounds such as pyridazine, pyrimidine, pyrazine, quinoline, isoquinoline, and acridine (Figure 1.35).

Unlike the nitrogen heterocycles, six-membered ring oxygen heterocycles cannot be aromatic unless the oxygen is positively charged (pyrilium ion). The reason is that there will always be at least two saturated

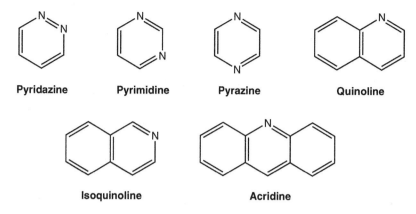

Pyridazine   **Pyrimidine**   **Pyrazine**   **Quinoline**

Isoquinoline   **Acridine**

Figure 1.35 Several aromatic six-membered heterocycles.

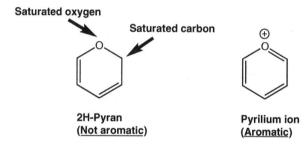

Saturated oxygen

Saturated carbon

2H-Pyran
(Not aromatic)

Pyrilium ion
(Aromatic)

Figure 1.36 Structural differences between 2H-pyran and the pyrilium ion.

(sp^3 hybridized) atoms in the ring. Why? Oxygen forms only two covalent bonds (unless it is charged). If oxygen is part of a ring, those two bonds join it to the remainder of the ring. If there are two C=C bonds in the ring, then there will be one other atom that does not have a double bond, and therefore, the requirement that each atom in the ring has a p-orbital cannot be met and the ring cannot be aromatic. While it is possible to prepare pyrilium ion, it is highly reactive and is generally employed as a reactive intermediate in organic synthesis (Figure 1.36).

Certain five-membered rings can also be aromatic. An example of this is the cyclopentadienyl anion. If cyclopentadiene is treated with a suitable base, a proton is removed from the sp^3 carbon and the electron pair from the C–H bond remains with that carbon. The hybridization of the carbon becomes sp^2 and the pair of electrons is contained within the p-orbital. At this point, the ring is flat, each atom in the ring has a p-orbital and there are six electrons in the π-system (four from double bonds and two from the filled p-orbital). The fact that this anion is

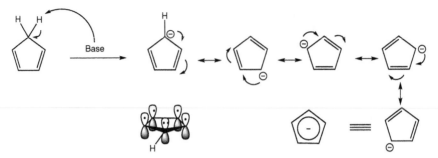

Figure 1.37 Aromaticity of the cyclopentadienyl anion. Note that cyclopentadiene itself is not aromatic because it has one saturated carbon.

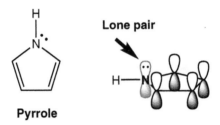

Pyrrole

Figure 1.38 Aromaticity in 1H-pyrrole. The lone pair on the nitrogen is located in the p-orbital and overlaps with the other p-orbitals to make the ring aromatic.

aromatic explains why cyclopentadiene is more acidic than most other hydrocarbons (Figure 1.37).

Fully unsaturated five-membered ring heterocycles may also be aromatic. 1H-Pyrrole, for example, is a five-membered nitrogen heterocycle with two double bonds. The nitrogen at first glance appears to be sp^3 hybridized because it is bonded to two carbons, a hydrogen, and also has a lone pair. The ring however is flat and has two double bonds (four electrons in a π-system). All that is needed to make the ring aromatic is one more p-orbital and two additional electrons. In nature, molecules tend to exist in their most stable (lowest energy) form. For 1H-pyrrole, this is achieved by making the ring aromatic. The nitrogen is actually sp^2 hybridized and the lone pair is housed within the p-orbital. These electrons join the π-system to complete the aromatic sextet (Figure 1.38).

A comparison of pyridine with pyrrole reveals an important point. Remember that the lone pair of pyridine is contained within an sp^2 orbital that is in the plane of the ring and that does not interact with the π-electron system. This lone pair is capable of forming bonds with another species that

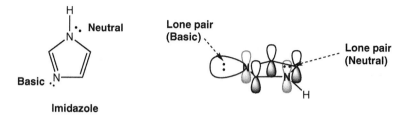

Figure 1.39 Structure of imidazole showing that one lone pair is part of the aromatic π-system, while the other is not.

has a vacant orbital, such as a proton. Thus, pyridine acts as a base. In pyrrole, however, the lone pair is located in a p-orbital that is part of the π-system making the ring aromatic. This lone pair is tied up and is therefore not capable of forming a bond with a proton. This means that unlike pyridine, pyrrole does not behave as a base.

Imidazole is another aromatic five-membered ring heterocycle that has two nitrogens. These nitrogens, however, have different properties. The one with its lone pair in a p-orbital uses those electrons to complete the aromatic sextet and this nitrogen, as in pyrrole, is neutral. The other nitrogen has its lone pair contained within an sp² orbital and those electrons are therefore in the plane of the ring, do not interact with the π-system, and can act as a base (Figure 1.39).

Several other aromatic nitrogen-containing five-membered ring heterocycles are shown below (Figure 1.40).

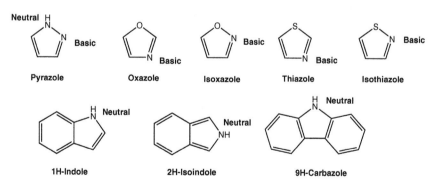

Figure 1.40 Several other five-membered and benzo-fused five-membered nitrogen-containing aromatic heterocycles.

Included in the structures shown above are several heterocycles containing oxygen or sulfur. These behave similar to pyrrole in that the oxygen or

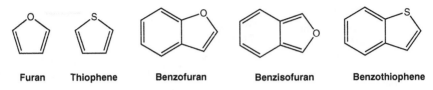

| Furan | Thiophene | Benzofuran | Benzisofuran | Benzothiophene |

Figure 1.41 A selection of five-membered oxygen and sulfur-containing aromatic heterocycles.

sulfur is sp^2 hybridized with one lone pair contained in the p-orbital completing the aromatic sextet. The remaining lone pair is in an sp^2 orbital, in the plane of the ring. Other aromatic five-membered ring heterocycles containing oxygen or sulfur are shown below (Figure 1.41).

REFERENCE

[1] L Pauling, *The Nature of the Chemical Bond*, third ed., Cornell University Press, Ithaca, 1960, pp. 111–144.

CHAPTER 2

The Three-Dimensional Structure of Organic Compounds

One of the critical factors associated with drug–receptor interactions is the three-dimensional structure of both the drug and receptor molecules. Efficient interaction between a drug and its receptor depends not only on the absolute and relative positioning of groups in space but also on the ability of the molecule to modify its shape by rotating about single bonds. These aspects of chemical structure are known as *stereochemistry* and *conformation* and are the topics covered in this chapter.

ISOMERS

Two or more compounds that have the same empirical formula but differ with respect to how the various atoms are joined are called *isomers*. Isomers are broken down into two broad categories. The first is *constitutional isomers*— compounds that differ in constitution or make-up. Thus cyclopropane and propylene (Figure 2.1) are constitutional isomers because cyclopropane is composed of three CH_2 (methylene) groups arranged into a three-membered ring, whereas propylene is acyclic with a vinyl and a methyl group. The alcohols *n*-propanol and *i*-propanol are also constitutional isomers that are similar chemically but differ in the position to which the hydroxyl group is attached (also called *positional isomers*). Thus *n*-propanol has the hydroxyl group attached to methylene with an ethyl group attached to the same carbon. In contrast, *i*-propanol has the OH group attached to a CH (methine) that is also attached to two methyl groups. Another example of this is seen with isobutylene and *cis*- or *trans*-2-butene. Isobutylene has one sp^2-hybridized carbon with two methyl groups and another one with two hydrogen atoms attached, whereas *cis*- and *trans*-2-butene have two sp^2 carbons each with one methyl group and one hydrogen atom. Thus, isobutylene and *cis*- or *trans*-2-butene are constitutional isomers. But what is the relationship between *cis*- and *trans*-2-butene? Since they have the same constitution but differ in the special arrangement of the various groups, they are called *stereoisomers*.

Organic Chemistry Concepts and Applications for Medicinal Chemistry
http://dx.doi.org/10.1016/B978-0-12-800739-6.00002-4

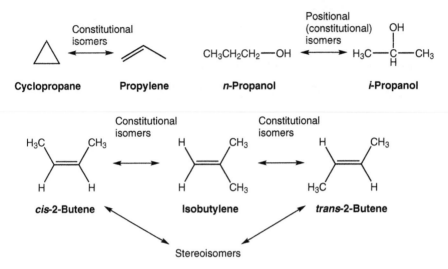

Figure 2.1 Examples of various types of isomers.

STEREOISOMERISM AT SATURATED CENTERS

In the preceding figure, *cis-* and *trans*-2-butene were shown to be stereoisomers by virtue of the spatial arrangement of groups around the double bonds. Stereoisomerism, however, can also occur at saturated centers. An example is illustrated in Figure 2.2. Structure **A** on the left has a carbon that is attached to H, OH, CH_2OH, and CHO. As drawn, the central carbon, CHO, and CH_2OH groups are in the plane of the paper, the H is behind the plane (depicted using a hashed wedge), and the OH is in front of the plane (solid wedge). If one were to set structure **A** in front of a mirror, the reflection would appear as structure **B**. Again, the central carbon, CHO, and CH_2OH groups would all lie in the plane of the paper, the H would be behind the plane, and the OH in front of the plane. To determine if structure **A** and **B** are identical, one would need to grab hold of the CHO group and rotate structure **B** by 180°. This would allow the CHO, central carbon, and CH_2OH groups of **B** to line up exactly as in structure **A**. In doing so however, one sees that the OH and H are exactly opposite in structure **B** to the way they are shown in **A**. Thus, **A** and **B** are mirror-image isomers that cannot be superimposed. Such stereoisomers are called *enantiomers.*

The central tetrahedral carbons of **A** and **B** are substituted by four different groups and therefore have no inherent symmetry. They are known as *asymmetric* or *chiral centers* and can exist in two enantiomeric forms. If

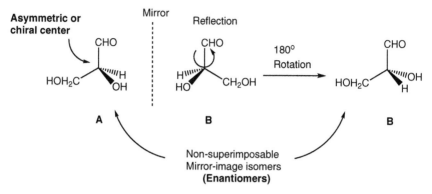

Figure 2.2 Mirror-image isomers of a saturated compound having a chiral center. Note that the two isomers cannot be superimposed (enantiomers).

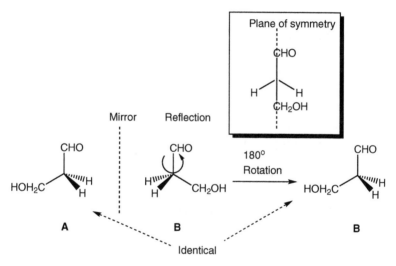

Figure 2.3 Mirror-image structures (**A** and **B**) of a saturated compound that possesses a plane of symmetry are completely superimposable, and represent the exact same compound.

however two groups attached to the central carbon are identical, then a plane of symmetry exists and enantiomeric structures are not possible (Figure 2.3).

Enantiomers have the ability to rotate the plane of polarized light that is passed through a solution of the compound. Because of this unique property, these compounds are said to be *optically active*. Optical rotation ($[\alpha]$) is a physical property of the compound that is measured using a *polarimeter* and is expressed in terms of degrees of rotation. Two compounds that are enantiomers have optical rotations that are of equal magnitude but opposite sign. When two enantiomers are present in equal amounts, the resulting mixture becomes optically inactive because the rotations cancel out. An equal mixture of two enantiomers is called a *racemic* mixture and is usually designated with a (±)- or DL- preceding the chemical name of the compound. With the exception of optical rotation, enantiomers have identical physical and chemical properties and are indistinguishable in an *achiral* environment.

While three-dimensional structure is most commonly depicted in two-dimensions using wedges and hashed bonds as shown above, there is another convention called *Fisher projections* that finds use mainly in carbohydrate and amino acid chemistry. In Fisher projections, intersecting horizontal and vertical lines take the place of wedged bonds. Groups attached to the vertical line project behind the plane of the paper while those on the horizontal axis project in front of the paper. A carbon atom is implied at the point where the lines intersect. Fisher projections are particularly good at highlighting the stereochemical relationships between different structures. The two enantiomers of glyceraldehyde are drawn below as Fisher projections. The designation (+)- or (−)- before the word glyceraldehydes indicates the sign of the optical rotation of that enantiomer. Notice that the two structures differ only with respect to the relative orientation of two groups. In general, switching any two groups of a Fisher projection generates the enantiomeric structure (Figure 2.4).

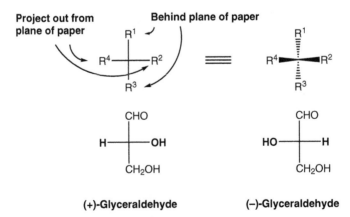

(+)-Glyceraldehyde   (−)-Glyceraldehyde

Figure 2.4 Top: A Fisher projection with the implied three-dimensional arrangement shown to the right. Bottom: The enantiomers of glyceraldehyde depicted as Fisher projections. Note that the two structures differ only by the arrangement of two groups (the H and OH).

Enantiomeric structures are also generated when a Fisher projection is rotated by 90° in either direction (Figure 2.5).

Two Fisher projections related by a 180° rotation represent the identical compound (Figure 2.6).

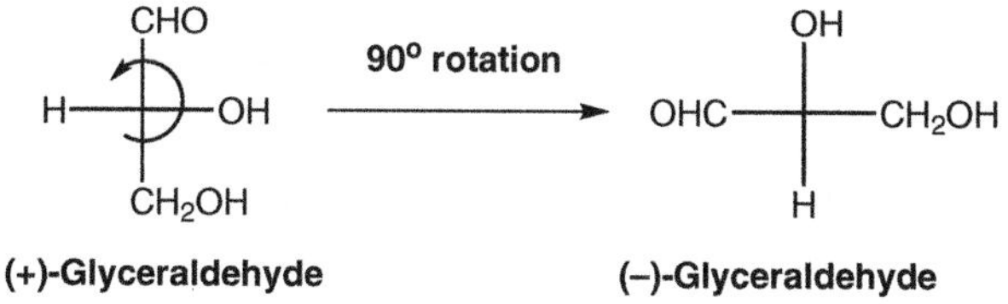

(+)-Glyceraldehyde   **(-)-Glyceraldehyde**

Figure 2.5 An example demonstrating that rotation of a Fisher projection in either direction by 90° generates an enantiomeric structure.

(+)-Glyceraldehyde   **(+)-Glyceraldehyde**

Identical

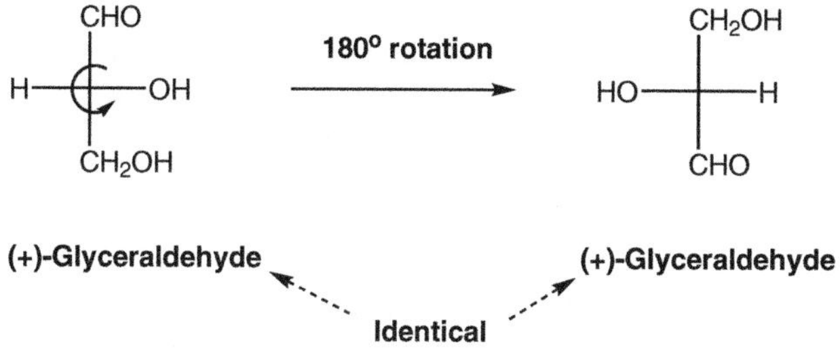

Figure 2.6 When two Fisher projections are related as 180° rotations, they represent the same structure.

Holding any one group constant and rotating the remaining three groups in either direction also gives rise to the identical structure (Figure 2.7).

The actual arrangement in space of the four groups surrounding a chiral center is known as the *absolute configuration* of that center. There are two main methods that are used for designating absolute configuration. The first

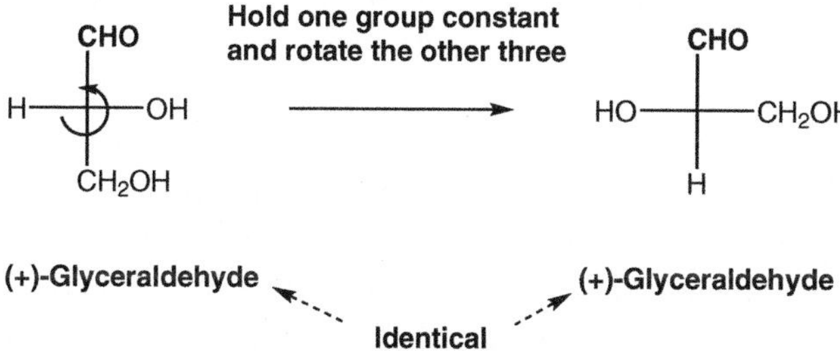

(+)-Glyceraldehyde   **(+)-Glyceraldehyde**

Identical

Figure 2.7 If one group is held constant and the other three are rotated in either direction, the resulting Fisher projection represents the same structure.

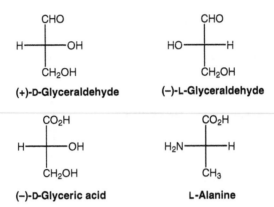

Figure 2.8 Using D- and L-glyceraldehyde to assign other chiral molecules as D or L. It is helpful to remember that L-glyceraldehyde has the OH group on the *left* in the Fisher projection when the aldehyde is at the top of the structure. The glyceric acid enantiomer shown has the OH on the right with the acid group on top and is assigned as D-glyceric acid, whereas the alanine isomer has the amino group on the left with the acid on top and so is L-alanine.

and older system relies on analogy with the randomly selected structures of the enantiomers of glyceraldehyde. The enantiomer labeled (+)-glyceraldehyde as shown in Figure 2.8 was given the designation D-glyceraldehyde (D from *dextrorotatory* or rotation to the right) while the (−)-enantiomer was designated as L-glyceraldehyde (L from *levorotatory* or rotation to the left). When a compound is drawn as a Fisher projection in the same relative orientation as one of the glyceraldehyde enantiomers, i.e. with the carbon of highest oxidation state at the top and other similar groups aligned accordingly, that enantiomer can be assigned as either D or L. Two examples are shown in Figure 2.8.

It is important to be aware that the designations D or L do not necessarily coincide with the sign of the optical rotation (except for glyceraldehyde) as evidenced by the glyceric acid enantiomer in the figure above. All of the naturally occurring amino acids (except glycine, which does not have a chiral center) have the L-configuration. Thus the alanine enantiomer shown above is L-alanine since it can be arranged so that the highest oxidation state carbon is on top, the NH_2 takes the place of the OH in glyceraldehyde, and the H is in the same location as in L-glyceraldehyde. When a compound has several chiral centers, it is drawn as a Fisher projection with the carbon of highest oxidation state at the top. The orientation of the chiral center that is most distant from it then determines whether it belongs to the D- or L-series (Figure 2.9).

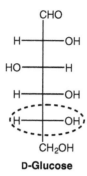

D-Glucose

Figure 2.9 This open-chain structure of glucose is depicted as a Fisher projection with the carbon of highest oxidation state (CHO) at the top. The most distant chiral center has the OH group to the right allowing this enantiomer to be assigned as D-glucose.

Unfortunately, a method that relies on analogy becomes much more difficult to apply as the compound in question diverges from the structure of the reference compound. For example, in attempting to assign the enantiomer of 2-chloro-2-phenylbutane shown below to either the D- or L-series which groups would be matched up with the CHO, CH$_2$OH, H, and OH of glyceraldehyde? There is no obvious answer. This then highlights the need for an alternative method of designating absolute configuration that does not rely upon analogy (Figure 2.10).

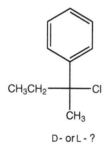

D- or L - ?

Figure 2.10 When the groups attached to a chiral center deviate greatly from those of glyceraldehydes, then assignment as D or L becomes difficult.

Such a method was developed by *Cahn, Ingold*, and *Prelog* and provides an unambiguous method for assigning absolute configuration at any chiral center [1]. The basis of this system is the assignment of priorities to the four groups attached to the chiral center with 1 representing the highest and 4 the lowest priority group. To assign priorities, a series of rules is applied as follows:

1. The atom with the higher atomic number has the higher priority. A lone pair of electrons (possible if the chiral center is an atom other than carbon) has the lowest possible priority.

2. An isotope with a higher atomic mass has priority over one with lower atomic mass, e.g. $^{14}C > ^{13}C > ^{12}C$ (regular carbon); ^{3}H (tritium, T) > ^{2}H (deuterium, D) > H.

For two atoms with the same priority, these rules are then applied to those atoms that are directly attached. For example, if a CH_3 and a CH_2OH group are attached to a chiral center, the directly attached atoms are both carbon. One carbon has three hydrogens attached, while the other has two hydrogens and an oxygen. The group with the oxygen therefore has higher priority. This test is applied continuously as you move away from the chiral center until a difference between substituents is observed. When a substituent includes a branch point, take the branch that will first show a difference between the two groups.

When the substituents are part of a double or triple bond, the following rules apply:

3. Double bonds count as two single bonds. Thus $CH_3C{=}O$ attached via the carbonyl carbon is counted as a carbon bonded to two oxygens and one carbon.

4. Triple bonds count as three single bonds. An $N{\equiv}C$ group counts as if carbon is bonded to three nitrogens.

5. Phenyl rings count as a series of alternating double and single bonds. For two substituted phenyl rings, substituents *ortho-* to the point of attachment have priority over *meta-*substituents, which have priority over *para-*substituents.

Once all four groups have been assigned a priority, then the chiral center is viewed from the side opposite to the group of lowest priority. If the order of priority of the remaining three groups decreases in a clockwise fashion, the chiral center has the *R-configuration* (from the Latin *rectus* for right). Alternatively, if the priorities decrease in a counterclockwise fashion, the chiral center has the *S-configuration* (from the Latin *sinister*, for left-handed). Shown below in Figure 2.11 is the use of the Cahn–Ingold–Prelog system for assigning absolute configurations to the glyceraldehyde enantiomers. In the example on the left, OH has the #1 priority because oxygen has the highest atomic number. Two substituents have a carbon bonded to the chiral center. The CHO group has carbon bonded to two oxygens (double bond) and one hydrogen, whereas in the CH_2OH group carbon is bonded to two hydrogens and one oxygen. Thus CHO has the second highest priority and CH_2OH the third highest. Since H has the lowest atomic number, it is the lowest priority (#4) group. If the structure is turned slightly so that H is behind the paper and the other three groups are in the plane, it becomes clear that going from #1 to #2 to #3 proceeds in a clockwise manner and this enantiomer is therefore *R*-glyceraldehyde. On the right, the priorities are the same but going from #1 to #2 to #3 proceeds counterclockwise and so this is *S*-glyceraldehyde. It is important to note that *R* and *S* do not necessarily correlate with D- and L- or (+)- or (−)- designations. It is not always convenient or possible to view a chiral center from the side opposite to the lowest priority group. In that case, priorities are assigned to the groups and the chiral center is viewed from the same side as the lowest priority group. The direction (clockwise or counterclockwise) of decrease of priorities is then observed and the opposite assignment is made (Figure 2.12).

Chiral centers that are drawn as Fisher projections can also readily be assigned using the Cahn–Ingold–Prelog system. If the group of lowest priority is on the vertical axis (behind the paper), the order of the decrease of priorities of the remaining groups is observed and the normal assignment

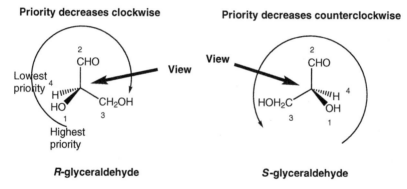

Figure 2.11 Using the Cahn–Ingold–Prelog method to assign absolute configuration at a chiral center.

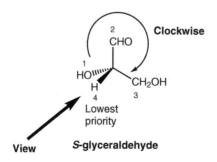

Figure 2.12 An example of making the opposite assignment (*R* for counterclockwise, *S* for clockwise decrease in priorities) when the lowest priority group is coming out of the plane of the paper.

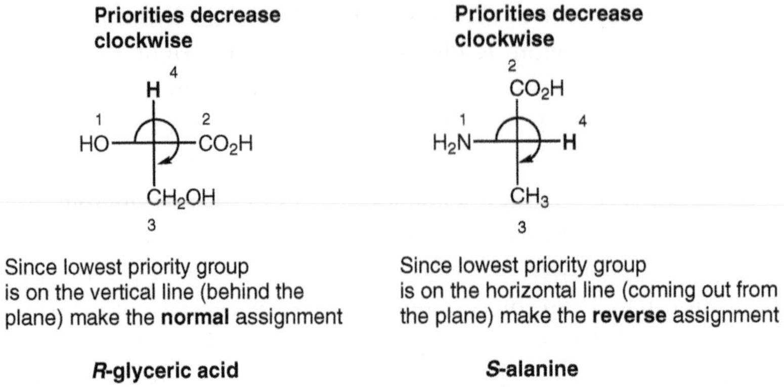

**Priorities decrease
clockwise**

**Priorities decrease
clockwise**

Since lowest priority group
is on the vertical line (behind the
plane) make the **normal** assignment

Since lowest priority group
is on the horizontal line (coming out from
the plane) make the **reverse** assignment

R-glyceric acid

S-alanine

Figure 2.13 Assigning chiral centers as R or S when drawn as Fisher projections.

(R for clockwise, etc.) is made. If, however, the lowest priority group is on the horizontal axis, the order of decrease in priorities is noted and the opposite assignment is made. Examples of this are shown above (Figure 2.13).

Compounds that have more than one chiral center may have a number of stereoisomers. For a compound with n chiral centers, a maximum of 2^n stereoisomers is possible. In the example below, the compound has two chiral centers and the structures of all four isomers are shown (Figure 2.14). The first isomer has the 2R,3R-configuration while the configuration of the second is 2S,3S. These compounds are enantiomers because the configuration at each chiral center is inverted. The third compound has the 2S,3R-configuration. Its relationship to the first two compounds is that of a diastereomer since only one chiral center is inverted relative to the other isomers. By definition, *diastereomers* are stereoisomers that are not enantiomers. Unlike enantiomers, diastereomers may have different chemical and physical properties in any environment. Their optical rotations (if they have any) may vary in both magnitude and sign. The fourth compound has the 2R,3S-configuration. This compound is the enantiomer of the compound with the 2S,3R-configuration because again, each chiral center is inverted. It is diastereomeric to the 2R,3R- and 2S,3S-isomers. In general, compounds with multiple chiral centers are related as enantiomers *if and only if* each and every chiral center is inverted from one structure to the next. Two diastereomers that differ by the configuration at only a single chiral center are called *epimers*.

Tartaric acid has two chiral centers and three stereoisomers. 2R,3R-Tartaric acid and 2S,3S-tartaric acid are enantiomers and are optically active. The third isomer, *meso*-tartaric acid, is not optically active. This molecule has a plane of symmetry that bisects the C2–C3 bond reflecting one

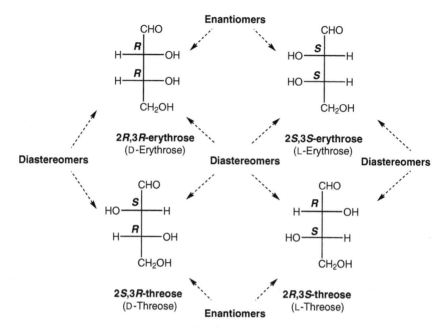

Figure 2.14 When a compound has more than one chiral center, an isomer in which each chiral center is inverted from the original structure will be an enantiomer. If even one chiral center retains the same configuration as the original structure, then they will be diastereomers.

end of the molecule onto the other. The presence of such a symmetry element rules out the possibility of an enantiomeric structure and renders the compound optically inactive. Whenever *meso*-isomers are formed, the total number of stereoisomers will be less than 2^n (Figure 2.15).

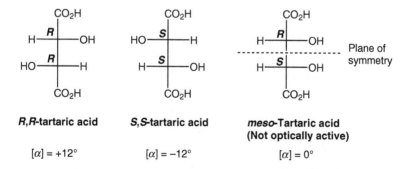

Figure 2.15 If a compound with multiple chiral centers has an isomer that has an internal plane of symmetry, that isomer will not be optically active, and is known as a *meso*-isomer. A *meso*-isomer cannot have an enantiomer and the presence of such isomers will decrease the total number of stereoisomers that are possible for *n* chiral centers from a maximum of 2^n.

PROCHIRALITY

Ethanol (CH_3CH_2OH) has no chiral centers. The two hydrogens of the CH_2 group are identical in an achiral environment. If however one H was replaced by a group X, and then in a second compound, the other H was instead replaced by X, the resulting structures would be enantiomers. These hydrogens are therefore referred to as *enantiotopic* hydrogens. In an achiral environment, enantiotopic groups are equivalent (Figure 2.16).

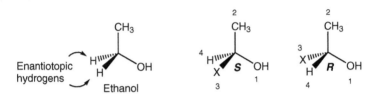

Figure 2.16 Ethanol is an achiral molecule but the two hydrogens on the methylene group are enantiotopic (prochiral) and may be differentiated in a chiral environment.

Enantiotopic groups may, however, be distinguished when placed into a chiral environment. For example, citric acid has two enantiotopic CH_2CO_2H groups. It is known that within the Kreb's cycle enzymatic oxidation of citric acid occurs at only one of these groups to give, after several further transformations, α-oxoglutaric acid. Enzymes, being highly chiral molecules, can distinguish between the enantiotopic groups of citric acid. Compounds that possess enantiotopic groups are said to be *prochiral*. The enantiotopic groups can be labeled either Pro-*S* or Pro-*R* by applying the priority rules and assigning a slightly higher priority to the group being labeled than its counterpart. In citric acid, it is the Pro-*S* CH_2CO_2H group that undergoes enzymatic oxidation. Another example of prochirality is the metabolic hydroxylation of only the Pro-*S* phenyl group of the antiepileptic drug phenytoin (Figure 2.17).

Just as groups can be enantiotopic, so too can the faces of an sp^2 center, such as a C=C or C=O bond. For example, reduction of 2-butanone can occur by delivery of hydride from the front face of the carbonyl group or from the rear. The resulting alcohols are enantiomers. The faces can be labeled using the priority rules. If the three groups attached to the sp^2 center decrease in priority in a clockwise fashion as you observe it, that face is labeled *Re*; if counterclockwise, the face is *Si*. If the sp^2 center is observed from the *Re*-face, the opposite face must be *Si*. While in an achiral environment, attack at either the *Re*- or *Si*-face is equally likely resulting in a racemic mixture of products. In a chiral environment, or if a chiral reagent was allowed to react, a single stereoisomer is likely to be the major product. Examples of

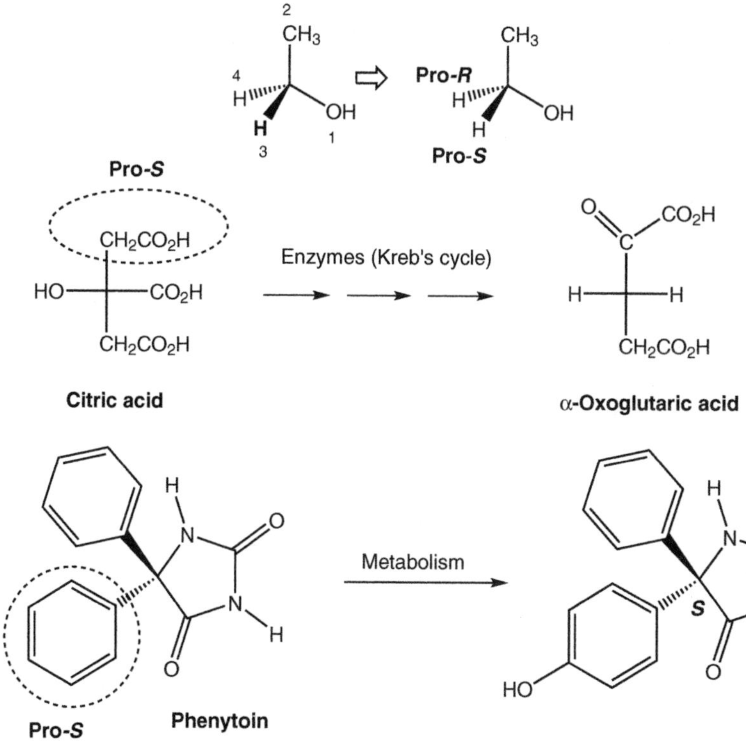

Figure 2.17 Top: The designation of enantiotopic protons as Pro-R and Pro-S. Middle: The Pro-S acetic acid moiety of citric acid is selectively oxidized in the chiral environment of the Kreb's cycle. Bottom: The antiepileptic drug phenytoin undergoes selective metabolic oxidation on the Pro-S phenyl group because metabolizing enzymes create a chiral environment.

enzymatic selectivity for enantiotopic faces are known. The anticoagulant (+)-R-warfarin undergoes metabolic reduction in the presence of NADPH from the Re-face of the keto-group to give as the major diastereomer the R,S-alcohol shown below (Figure 2.18).

STEREOISOMERISM AT UNSATURATED CENTERS

In the beginning of this chapter, the compounds cis- and trans-2-butene were used as examples of stereoisomers. While these familiar designations are adequate to describe the geometrical relationship of a disubstituted alkene with cis- referring to groups on the adjacent unsaturated carbons being on the same side of the double bond, while trans- signifies that the two groups are on opposite sides. The situation is not always so simple however.

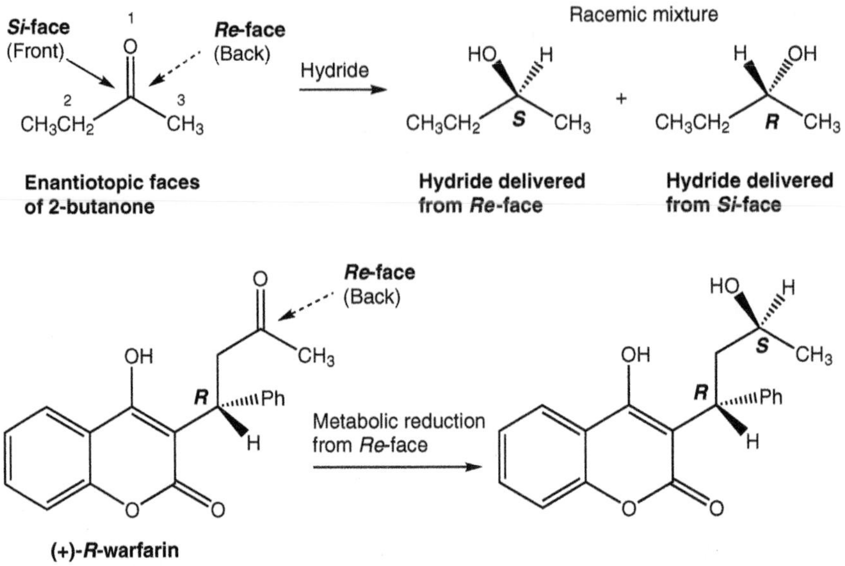

Figure 2.18 Top: Unsaturated centers that are not symmetrically substituted have enantiotopic faces called *Re* and *Si*. In an achiral environment, such as reduction by a nonchiral hydride reagent, the enantiotopic faces will be attacked equally giving rise to a racemic mixture of products. Bottom: The anticoagulant warfarin undergoes metabolic reduction selectively from the *Re*-face of the ketone.

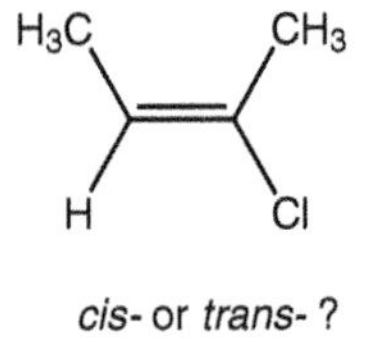

cis- or *trans*- ?

Figure 2.19 While *cis* and *trans* are adequate to denote the geometrical isomers of simple olefins such as 2-butene, this becomes more difficult as the degree of substitution around the double bond increases.

Consider the compound shown in Figure 2.19. Should this be named *cis*-2-chloro-2-butene or *trans*-2-chloro-2-butene? To answer this question, you need a point of reference. If you are referring to the methyl groups, then *cis*-would be correct. If however your point of reference is the larger group on each carbon, then it would be *trans*- since chlorine as a third-row element is larger than carbon. To alleviate such ambiguity, an alternative system for designating the geometry at double bonds has been devised that works for all double bonds. This system employs the same Cahn–Ingold–Prelog priorities that were employed for assigning *R*-/*S*-stereochemistry at saturated centers. In this case, the group with the larger priority on each end of the double bond is identified. The geometry is designated as Z- (from the German word *zusammen*, meaning together) when the groups of higher priority lie at the same side of the double bond. When the groups of higher priority are on opposite sides of the double bond, the geometry is E- (from the German word *entgegen*, meaning opposite). Thus in the example shown above, the highest priority group on the left side of the bond is the methyl, while it is the chlorine on the other end. Since they are on opposite sides of the double bond, the compound is E-2-chloro-2-bu-tene. Several other examples are shown below. Note that this system also works with heteroatom-containing double bonds such as C=N, N=N, etc. A carbonyl (C=O) cannot have E/Z-geometry because the oxygen has two equivalent lone pairs (Figure 2.20).

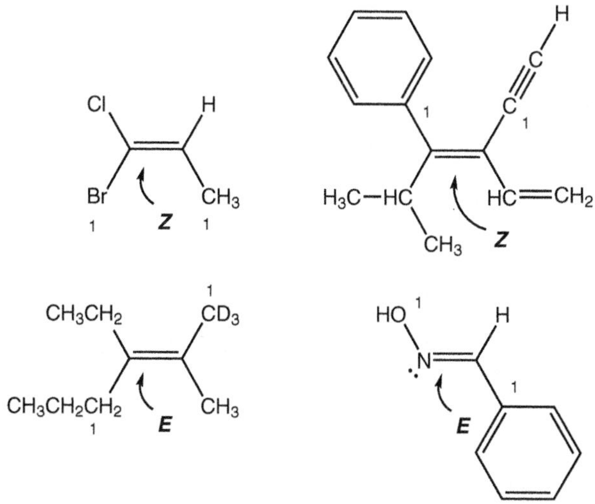

Figure 2.20 Several examples of double bonds assigned using the E/Z designation. Note that terminal olefins (those with =CH$_2$) and double bonds in rings smaller than eight-membered (including those in benzene rings) are not assigned.

In compounds having two or more double bonds, the geometry of each must be specified in the name. An example of this is shown below with the compound Z,E-octa-2,5-diene. The Z-designation belongs to the double bond starting at the 2-position and the E- to that starting at the 5-position (Figure 2.21).

While old habits are hard to break and people still refer to the butene examples at the beginning of the chapter as *cis* and *trans*, the use of E and Z is preferred. Does that mean that we have no more use for *cis* and *trans*? As

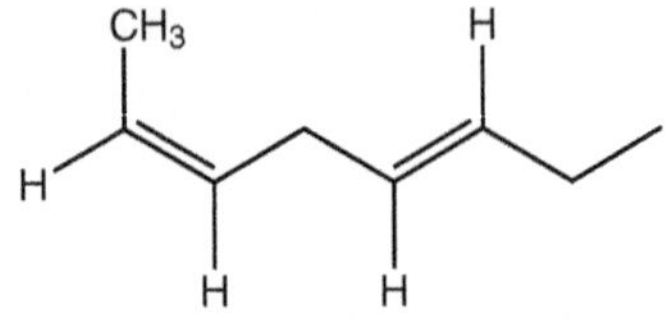

Z,E-octa-2,5-diene

Figure 2.21 When a compound has more than one double bond, each must be assigned as being either *E* or *Z*.

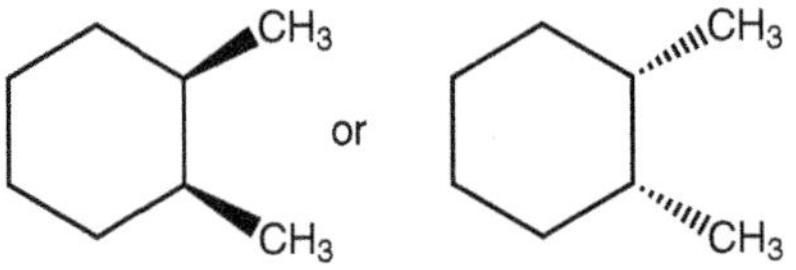

or

cis-1,2-Dimethylcyclohexane

or

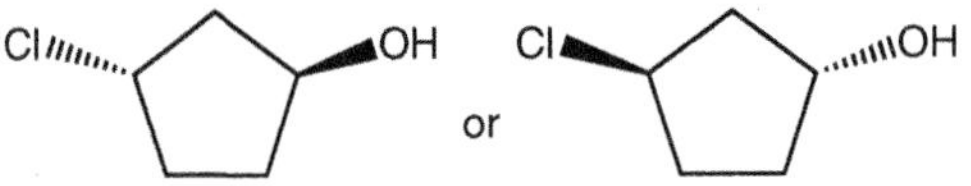

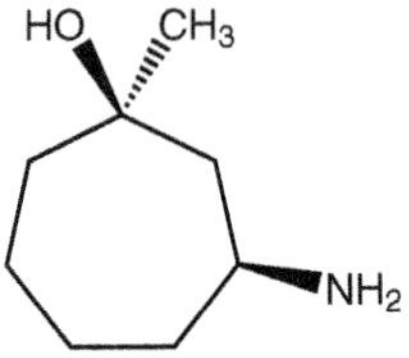

trans-3-Chlorocyclopentanol   **cis-3-Amino-1-methylcycloheptanol**

Figure 2.22 Left: Two substituents attached to a saturated ring are designated with the relative stereochemical indicator *cis* if both are on the same face of the ring and *trans* if on opposite faces. Right: When there are two substituents attached to the same ring carbon, then a substituent located elsewhere on the ring is *cis* if it is on the same face as the substituent of higher priority.

it turns out, these terms are used to describe the geometrical relationship of two groups attached to a saturated ring. Thus *cis*-1,2-dimethylcyclohexane has methyl groups on adjacent carbons of the ring that are either both above or below the ring. This is an example of relative stereochemistry. In *trans*-1-chlorocyclopentan-3-ol, the chlorine can be above the plane of the ring if the alcohol group is below it, or visa versa. In the absence of an absolute stereochemical designator such as *R* or *S*, it does not matter which is up and which is down. If there are two different substituents on one ring carbon and another substituent on another ring carbon, then the *cis-/trans*-designation refers to the higher priority group on each substituted ring carbon (Figure 2.22).

The designations *cis* and *trans* are also used to describe the stereochemistry of fused saturated rings at the bridgehead positions. Notice the difference between *trans*-decalin and *cis*-decalin with regards to the overall

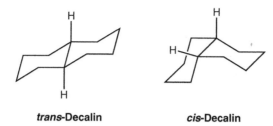

trans-Decalin   **cis-Decalin**

Figure 2.23 For fused saturated rings, groups that are external to the rings at the bridgehead positions (atoms that are common to both rings) are designated *cis* if they are both on the same face and *trans* if they are on opposite faces.

planarity of the ring system. *Trans*-decalin with the bridgehead hydrogens on opposite faces of the ring system is nearly planar, whereas *cis*-decalin has a sharp bend in the molecule where the rings are joined (Figure 2.23).

CONFORMATIONS

Two atoms that are joined by only a single bond can rotate independent of one another about that bond, much as two wheels connected by an axle. The different arrangements that result from these rotations are called *conformations*. Unlike stereoisomers, conformations arise without the need for breaking or forming of any new bonds. As the atoms rotate, any groups that are attached to them will become either closer together or further away from each other in space. When groups approach each other, the potential energy rises in part because the space occupied by each group begins to overlap. The larger the groups, the more they overlap, and the higher the energy of that particular conformation. While an infinite number of conformations is possible by rotation about a single bond, only those that correspond to an energy maxima or minima will be discussed. If for example we have two sp^3-hybridized carbons joined by a σ-bond and one of the carbons is held constant while the other is rotated through 360°, we can create a plot of energy vs *dihedral angle, θ*. In the figure below, if we let atom **A** and carbons **B** and **C** define a plane, then the dihedral angle is defined as the deviation of atom **D** from that plane (Figure 2.24).

A good illustration is provided by the simple hydrocarbon ethane. Rotation about the C–C bond results in two distinct conformations, one in which all the hydrogens on one carbon line up directly behind those of the second carbon, and one where the hydrogens on one carbon fall directly between two hydrogens on the second carbon. The first of these

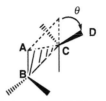

Figure 2.24 A pictorial representation of dihedral angle. Atoms **A**, **B**, and **C** define a plane. The dihedral angle (θ) is the deviation of atom **D** from that plane.

conformations is called *eclipsed* while the second is *staggered*. The eclipsed conformation is higher in energy than the staggered conformation. This increase in energy is associated with repulsive forces created as the distance between hydrogens approaches the sum of their atomic radii. These repulsive forces are known as *nonbonded interactions*. Repulsion between electrons in the filled C–H σ-orbitals is also greatest when those bonds are eclipsed, which contributes to the increased energy of such conformations. In the low energy or staggered conformation, nonbonded interactions are minimized. A plot of dihedral angle vs relative energy is shown below. The carbon to the front is held steady and the rear carbon is rotated; one hydrogen is shown in bold font to allow the reader to follow its rotation. Notice that there is an approximately 3 kcal/mol difference in energy between the eclipsed and staggered conformations, with each eclipsing interactions between hydrogens contributing 1 kcal/mol to the energy (Figure 2.25). It is useful to remember that a 3 kcal/mol difference in energy between the two conformations corresponds to 99.999% staggered and 0.001% eclipsed at room temperature.

The replacement of one hydrogen on each of ethane's carbons with a methyl group generates *n*-butane. This complicates the conformational analysis of *n*-butane if one rotates about the central C–C bond giving rise to four distinct conformations shown below. If one methyl group and the next two carbons define a plane, then the dihedral angle will be determined by the deviation of the remaining methyl group from that plane. When $\theta = 0°$, two methyl groups eclipse each other and there are also two pairs of eclipsing hydrogens. This arrangement, called a *fully eclipsed* conformation, is highest in energy. At $\theta = 60°$, the hydrogens are staggered and the methyl groups are 60° apart. This represents a lower energy arrangement and is called a *gauche* conformation. When the dihedral angle reaches 120°, there are two methyl groups eclipsing hydrogens and one H–H eclipsing interaction. Such a conformation will be higher in energy than gauche, but less

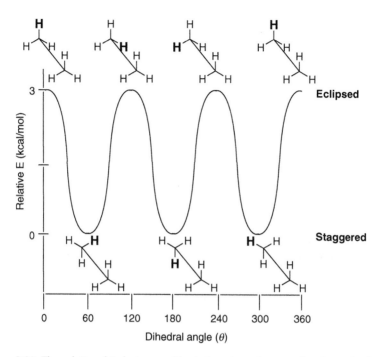

Figure 2.25 The relationship between dihedral angle and energy in ethane. In this figure, the hydrogen on top front and the two carbons define a plane and the dihedral angle is the deviation of the bolded hydrogen from that plane. The back carbon is rotated through 360° with an energy maxima occurring whenever the hydrogens fall directly behind those in front (eclipsed). The energy is minimized when the back hydrogens fall directly between those in front (staggered).

than fully eclipsed and is called *partly eclipsed*. A further 60° rotation results in a 180° dihedral angle and puts the two methyl groups as far away from each other as possible. Since all groups are staggered, this is the lowest energy conformation of butane and is called *anti*. If a statistical sampling of *n*-butane molecules at room temperature were taken, it would be found that most are in the lowest energy anticonformation, with the majority of the others being gauche. The high–energy conformations are generally not highly populated but serve as intermediate structures between low–energy conformations (Figure 2.26).

Cyclic compounds in general have fewer available conformations than acyclic compounds. Cyclohexane, for example, has four conformations called the *chair, half-chair, twist-boat,* and *boat*. Of these, the chair conformation is the lowest in energy (by about 5 kcal/mol) because of the absence of eclipsing interactions. In the boat conformation, there are four sets of

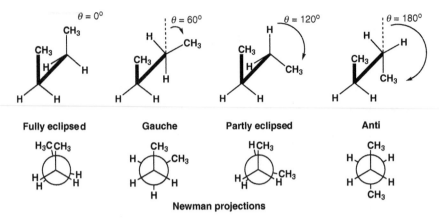

Fully eclipsed   **Gauche**   **Partly eclipsed**   **Anti**

Newman projections

Figure 2.26 The four distinct conformations of *n*-butane shown as sawhorse projections (top) and Newman projections (bottom). In a Newman projection, one is looking down a C–C bond with the back carbon represented by the circle and with bonds originating in the center of the circle being attached to the closer carbon. It is a common practice to skew the groups attached to the back carbon slightly in eclipsed conformations to make the attached groups easier to see.

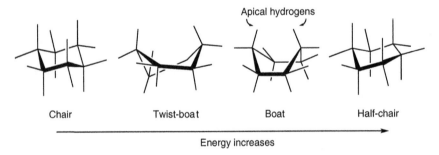

Chair   Twist-boat   Boat   Half-chair

Energy increases

Figure 2.27 The four conformations of a cyclohexane ring shown in order of increasing energy. Cyclohexanes prefer to exist as chair conformations whenever possible.

eclipsing interactions. In addition, one set of hydrogens on the apical carbons points directly at each other. These factors raise the energy of the boat conformation relative to the chair (Figure 2.27).

Examination of cyclohexane in the chair conformation reveals two distinct sets of hydrogens. Those that lie approximately in the plane of the ring are called *equatorial*. The other set of hydrogens is split evenly with three hydrogens above the plane of the carbon skeleton and three below. These are called *axial* hydrogens. Axial and equatorial positions equilibrate through a process known as *chair–chair interconversion*. This occurs rapidly at room

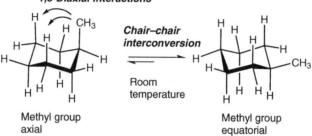

Equatorial
hydrogens

Axial
hydrogens

1,3-Diaxial interactions

*Chair–chair
interconversion*

Room
temperature

Methyl group
axial

Methyl group
equatorial

Figure 2.28 Top: Cyclohexane has two distinct groups of hydrogens. The equatorial positions are located roughly in the "plane" of the ring. Three axial positions are located on the top face and three on the bottom. Bottom: Chair conformations interchange rapidly at room temperature. In the process, groups that occupy an axial position in one chair become equatorial when the chair inverts. Axial substituents experience non-bonded interactions with axial protons located three carbons away that increase with the size of the group.

temperature. When a group larger than hydrogen occupies an axial position, nonbonded interactions occur with the axial hydrogens at the 3- and 5-positions. These interactions increase with the size of the substituent. For very bulky groups such as *tert*-butyl, these interactions are so severe that the energy barrier for chair–chair interconversion is not easily attained at room temperature and the group therefore remains largely equatorial. In general, conformations in which the larger substituents are equatorial are lower in energy than those in which such groups are axial (Figure 2.28).

Five-membered rings have two major conformations, called the *envelope* and the *half-chair*. The envelope conformation has four planar carbons with the remaining carbon situated above or below the plane, like the flap of an envelope. Each carbon in the ring can assume the flap position. The half-chair conformation has three planar carbons, another above the plane, and one below the plane. These conformations are very close in energy and interconvert rapidly at room temperature (Figure 2.29).

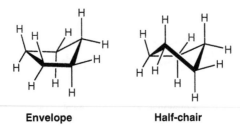

Envelope   **Half-chair**

Figure 2.29 The conformations of cyclopentane. Each carbon can assume the "flap" position of the envelope conformation.

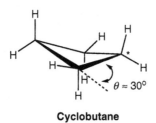

Cyclobutane

Figure 2.30 Cyclobutane exists mainly as a single conformation, which is puckered across the middle.

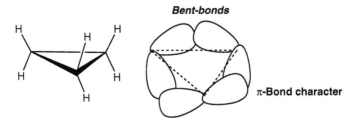

Figure 2.31 Since cyclopropane has only three carbons, they must be coplanar. This requires that the three hydrogens on the top face and those on the bottom face must be eclipsing.

Four-membered rings are puckered, with three carbons in a plane and the fourth above or below the plane by about 30 degrees. Unlike five- and six-membered rings, there are large deviations from the ideal bond angle (109.5°) for an sp^3 carbon in cyclobutane (88° for the C–C–C bond angle). This is manifested as *small-angle strain* energy and equals approximately 26 kcal/mol for cyclobutane (Figure 2.30).

Three-membered rings must be planar. The C–C–C bond angle of cyclopropane is approximately 60°. This results in 28.1 kcal/mol of small-angle strain energy. Both three- and four-membered rings undergo chemical reactions that can open the ring to relieve this strain energy. With most σ-bonds, the electron density is directed along the interatomic axis. Cyclopropanes, however, have the region of the greatest electron density for the C–C bonds diverted outward from this axis because of the severe bond angle. These bonds, called *bent* or *banana bonds*, have a lot of π-bond character. Reagents that react with double bonds will often react with the C–C bonds of a cyclopropane (Figure 2.31).

REFERENCE

[1] R.S. Cahn, C. Ingold, V. Prelog, Specification of molecular chirality, Angewandte Chemie, International Edition in English 5 (1966) 385–415.

CHAPTER 3

Functional Groups

Organic compounds are often comprised of a basic structure (a ring system or an open-chain array of carbon atoms) to which various groups (*substituents*) are attached. These groups affect the physical and chemical properties of the compound. Substituents with π-bonds or heteroatoms are called *functional groups* because they are capable of forming bonds (such as hydrogen bonds) and undergoing chemical reaction. Functional groups can also be part of the basic structure of the molecule. Pharmaceuticals are often richly endowed with functional groups that serve as sites for interacting with receptors, metabolic conversion, and for determining acid/base and partitioning properties. An example is seen below in the sedative diazepam. This compound has several functional groups including a lactam, an imine, two phenyl rings, and a chloro group (Figure 3.1).

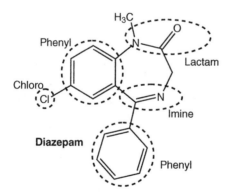

Figure 3.1 The chemical structure of the sedative drug Diazepam (Valium®) showing the presence of a variety of different functional groups.

COMMON FUNCTIONAL GROUPS

Some functional groups contain just a single atom, such as the chloro group shown above. Others are made up of two or more atoms working together as a unit (as with the lactam and the imine). A large number of functional groups (including lactams) contain the carbonyl group (C=O). It is important to be able to recognize common functional groups within larger molecules because that will provide a clue as to the types of reactions that the compound in question can undergo. The names of carbonyl-containing functional groups depend on the nature of the two groups that are attached to the carbonyl carbon. Some of the more common carbonyl-based functional groups are shown in Figure 3.2.

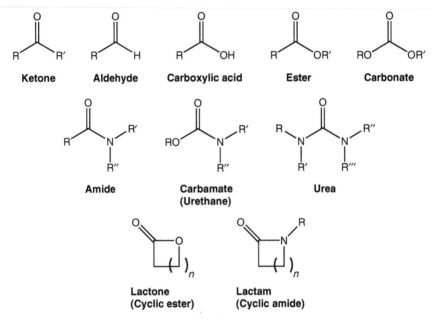

Figure 3.2 General structures of the common carbonyl-based functional groups. The symbols R, R', R" and R" represent any aliphatic, aromatic, alkenyl or alkynyl group.

Another set of functional groups in which nitrogen is the central atom is shown in Figure 3.3. Amines are often classified by the number of non-hydrogen substituents attached to the nitrogen. An RNH_2 group is called a *primary amine*, while RNHR' is a *secondary amine*, and RNR'R" is a *tertiary amine*. Amines always have a lone pair of electrons attached to the nitrogen. When that lone pair is used to form a fourth bond, the nitrogen acquires a positive charge. If one of the four bonds is made to hydrogen, then the resulting functional group is called *ammonium*. As will be seen in Chapter 4, the charge on nitrogen is not permanent since the hydrogen can still depart to reestablish the lone pair. Thus ammonium groups are acidic. If however all four bonds from nitrogen are made to carbon-based groups, this is known as a *quaternary ammonium* group and the charge on nitrogen is permanent because C–N bonds are not easily broken. In medicinal chemistry, this is a

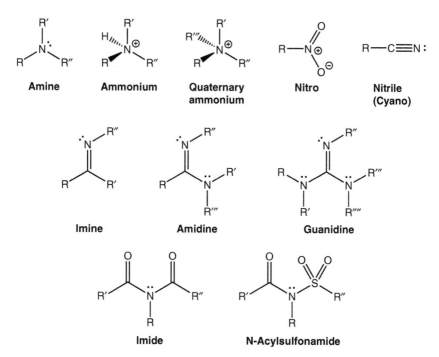

Figure 3.3 The general structures of several common nitrogen-based functional groups.

very important distinction because compounds with permanent charges have difficulty passing across certain membranes in the body. Also note that a nitro group (usually written as NO_2) always has a positive charge on the nitrogen, but this is offset by a negatively charged oxygen. The negative charge can be moved by resonance from one oxygen atom to the other.

Several functional groups that have a central sulfur atom are shown in Figure 3.4. Since sulfur is a third-row element, it is capable of existing in three different oxidation states, the lowest having sulfur bonded to two other elements and with two lone pairs associated with the sulfur. In the second highest oxidation state, sulfur forms four bonds and retains one lone pair. At its highest oxidation state, sulfur forms six bonds and has no lone pairs. The sulfur atom at all three oxidation states is tetrahedral. Thus a sulf-oxide with two different R groups can be a chiral center, with the oxygen and lone pair serving as the remaining two groups. Thiols are distinguished by having extremely unpleasant odors. The parent compound, hydrogen sulfide, H_2S, smells like rotten eggs. *tert*-Butylthiol is added to odorless natural gas in minute concentrations so that a gas leak can be detected by odor.

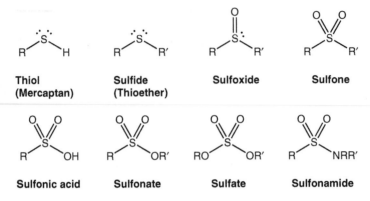

Figure 3.4 Some of the more common sulfur-based functional groups.

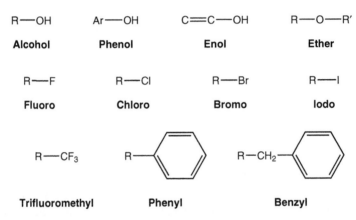

Figure 3.5 Several miscellaneous functional groups commonly found in drug molecules.

Some functional groups that do not fit into the previous categories are shown in Figure 3.5. The first three include a hydroxyl group (OH) but behave differently with respect to chemical properties. Each is a polar group that can form or accept hydrogen bonds. Alcohols behave as very weak acids but phenols and enols are stronger acids. Enols may revert to their more stable form, which has a carbonyl group. This requires that the hydrogen attached to the oxygen migrate to the far end of the double bond in a process called *tautomerization* (Figure 3.6). This establishes an equilibrium between the *enol* and *keto* tautomers, which usually favors the keto-form as shown for acetone. If there are structural features that stabilize the enol form, then it can be the major tautomer as shown for acetylacetone. In the enol form, the OH group can form an intramolecular hydrogen bond with

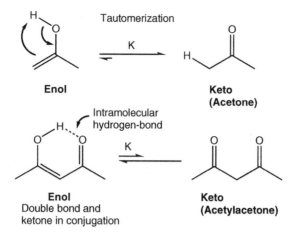

Figure 3.6 Top: Keto-enol tautomerization in acetone. The equilibrium for this molecule strongly favors the keto-tautomer. Bottom: The tautomeric equilibrium for 2,4-pentanedione (acetylacetone) favors the enol form because the double bond is conjugated with the remaining carbonyl group and the hydrogen of the enol can form a hydrogen bond with the carbonyl oxygen that creates a very stable six-membered ring.

the other ketone, which is especially stable because this arrangement forms a six-membered "ring". Also, the double bond of the enol is in conjugation with the carbonyl group. Both factors contribute to the enol form being the major tautomer for this compound.

Several other types of tautomerization are shown in Figure 3.7. In each case, the equilibrium lies well to the right. For keto-phenol tautomers, the favored phenol form is aromatic whereas the keto form is not. Enamines that have at least one hydrogen on the nitrogen are generally not stable and revert to the imine form. Aliphatic nitroso compounds with a hydrogen attached to the adjacent carbon are also usually not stable and exist primarily as oximes.

THE ELECTRONIC EFFECTS OF FUNCTIONAL GROUPS

Functional groups can be broadly placed into two categories according to the electronic effect they exert on the structure to which they are attached. Those that serve to remove excess electron density from the molecule are *electron-withdrawing* groups whereas those that can contribute electron density to an electron-deficient molecule are *electron-donating* groups. Electron-withdrawing groups do not actually remove excess electron density but they provide a mechanism for delocalizing it, while donating groups do the same for developing positive charges. The electronic properties of functional groups are the

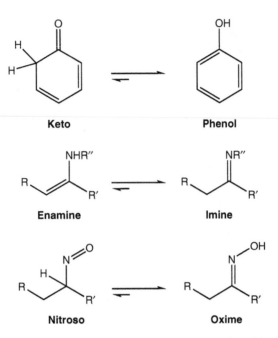

Figure 3.7 Several other examples of tautomerization. Top: Phenol-keto tautomerization strongly favors the phenol form because it is aromatic. Middle: Imine-enamine tautomerization favors the imine form unless there is no hydrogen on the nitrogen. Bottom: Oxime-nitroso tautomerization favors the oxime form. Aliphatic nitroso compounds are not stable unless the carbon with the N=O group has no hydrogen.

result of three distinct effects: inductive, field, and resonance. *Inductive effects* are created by the polarization of bonds due to electronegativity differences between atoms. For example, a bond between carbon and a more electronegative element such as oxygen is polarized such that the electron density in the σ-bond is drawn closer to the oxygen leaving a slight positive charge on the carbon. If this carbon is bonded to another carbon, then that bond is now polarized, with the electrons drawn slightly toward the positively charged carbon. This effect can be propagated through several bonds, although it decreases in intensity rapidly as the distance from the electronegative element increases. Thus inductive effects are transmitted through the σ-bonding framework of a molecule.

Polarization that occurs through space or is transmitted by solvent molecules rather than through bonds is known as a *field effect*. This effect, like the inductive effect, drops off substantially with distance. A good example of a field effect is seen with the two bicyclic carboxylic acids shown in Figure 3.8. The chlorines in each isomer are exactly the same number of

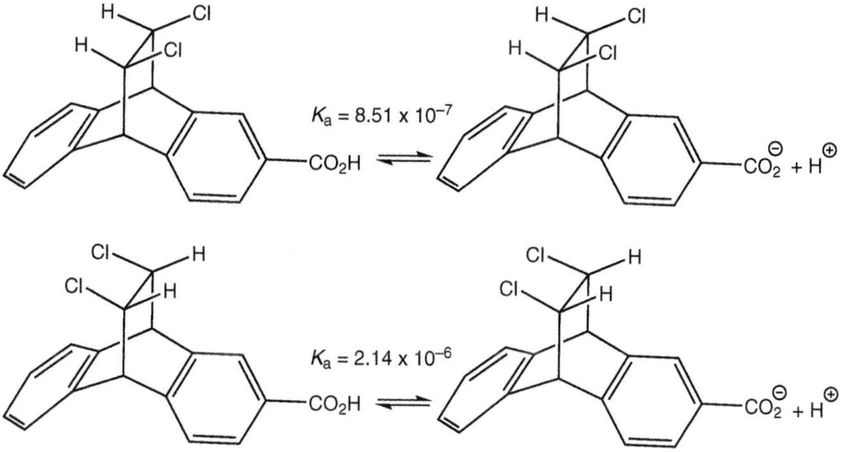

Figure 3.8 An example of a field effect. Note that the only difference between the top and bottom compound is the orientation of the two chlorine atoms with respect to the carboxylic acid group [1].

bonds away from the carboxylic acid suggesting that any inductive effects should be identical. The only difference between them is the distance in space from the chlorines to the carboxyl group. Field effects therefore account for the difference in acidity between these isomers [1]. While this example is an exception, it is often difficult to distinguish between inductive and field effects and they are often combined and are known collectively as *polar effects*.

Resonance effects have been discussed earlier and involve the movement of electrons through a conjugated system. Unlike field and inductive effects, resonance effects decrease much more slowly with increasing distance. These effects often contribute more heavily to the overall electronic effects of functional groups than do polar effects.

Now that we have some understanding of the origin of electronic effects in functional groups, we can use that knowledge to make assessments of which ones are electron withdrawing and which are donating. Consider the carbonyl-based functional groups shown in Figure 3.2. Since oxygen is considerably more electronegative than carbon, we expect polar effects to induce some negative charge on the oxygen and a partial positive charge on carbon. That positive charge will attract electron density from the molecule to which it is attached and it is expected that carbonyls will serve as electron-withdrawing groups. A carbonyl group also possesses a π-bond and so any source of electron density to which it is attached can be delocalized by resonance onto the oxygen. Thus the polar effects and resonance work in concert to make carbonyl-based functional groups good electron withdrawers (Figure 3.9).

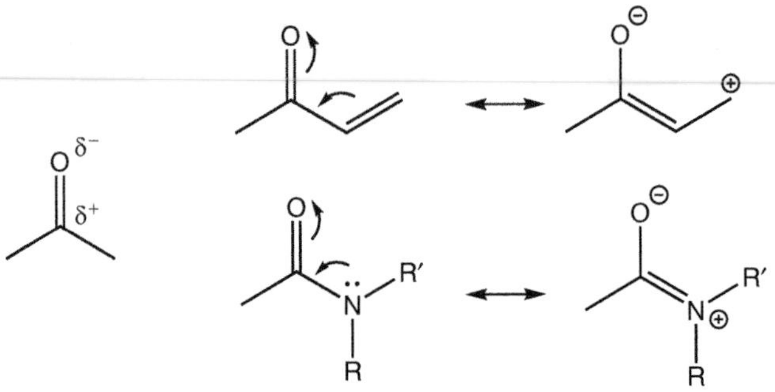

Figure 3.9 Left: Polar effects in a carbonyl group induce a partial positive charge on carbon and a partial negative charge on oxygen. Right: The carbonyl group withdraws electrons by resonance.

Nitro is a powerful electron-withdrawing group. Since the nitrogen always carries a positive charge, it exerts a strong polar effect on the compound to which it is attached. The presence of a double-bonded oxygen provides an alternative pathway for stabilizing negative charge by resonance. These effects can be seen using 4-nitrophenol as an example (Figure 3.10).

Nitriles are also strong electron-withdrawing groups that can stabilize excess electron density by both resonance and polar effects. The electronegativity difference between carbon and nitrogen induces a partial positive

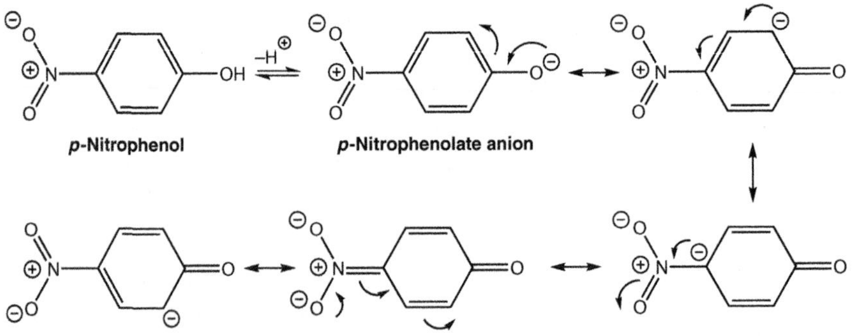

p-Nitrophenol   *p*-Nitrophenolate anion

Figure 3.10 The nitro group is one of the most powerful electron-withdrawing groups. The nitrogen always carries a positive charge that is balanced by a negatively charged oxygen, making the nitro group electrically neutral overall. In this example, we note the ability of nitro to withdraw electrons in an extended conjugated system by resonance.

charge on the carbon and the multiple bonds allow for the movement of electrons onto the nitrogen by resonance (Figure 3.11).

Figure 3.11 A nitrile carries a partial positive charge on the carbon due to polar effects and can also withdraw electrons by resonance.

Functional groups that have either one sulfur–oxygen double bond (sulfinyl; such as in sulfoxides) or two (sulfonyl; e.g. sulfones) are also strongly electron withdrawing. Sulfur resides one row below oxygen on the Periodic Table and is therefore a much larger element. This size difference reduces the efficiency of orbital overlap between the two atoms. As a result, whenever there is an S=O bond, there is also a very significant resonance structure in which S and O are joined by only a single bond, but with a formal positive charge on the less electronegative sulfur and a negative charge on oxygen. This induces a significant polar effect attracting electrons toward the sulfur. The double bond also allows for the migration of electrons onto the oxygen(s) by resonance. Functional groups with a sulfonyl unit have one additional resonance structure than those having a sulfinyl moiety, and are slightly stronger electron-withdrawing groups (Figure 3.12).

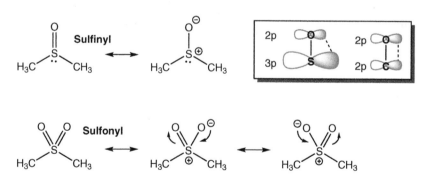

Figure 3.12 Top: A single S=O bond is called sulfinyl. Oxygen, being more electronegative than sulfur, carries a partial negative charge making the sulfur partially positive. The shadow box demonstrates why 2p orbitals on carbon and oxygen overlap efficiently to form carbonyl bonds, while the difference in size between the sulfur 3p and oxygen 2p orbitals makes S=O bond formation less efficient. Bottom: When sulfur has two double-bonded oxygens, the resulting group is called sulfonyl. Again there will be significant positive charge on sulfur and negative charge on oxygen that can be placed onto the other oxygen by resonance.

The halogens are all more electronegative than carbon and behave as electron-withdrawing groups by virtue of their polar effects. While it might be expected that fluorine, which is the most electronegative element, would be the strongest electron withdrawer of the halogens, this is not the case. While fluorine withdraws electron density strongly by its polar effects, it donates some of its electron density (remember all of the halogens have three lone pairs) back to carbon by resonance. This is possible because F and C are from the same row of the Periodic Table, and are thus of similar size, meaning their orbitals are also similar in size and can overlap efficiently to form a π-bond. But how could such a π-bond form to a saturated carbon? Remember that when hybrid orbitals from two atoms overlap to form a new σ-bond, an antibonding σ* orbital is also formed, although it is vacant. An orbital containing one of the fluorine's lone pairs can overlap with a vacant σ* orbital on carbon which allows electron density to flow from fluorine back to carbon. Thus while the polar effect of fluorine is electron withdrawing, the resonance effect is donating. Because of fluorine's very high electronegativity, the polar effect wins out and fluorine behaves as an electron-withdrawing group, but a relatively weak one. The other halogens are all much larger than carbon and back-donation by the mechanism employed by fluorine is inefficient. Therefore, chlorine, which is the most electronegative of the remaining halogens, is the strongest in terms of its electron-withdrawing potential. Bromine and iodine behave as weak electron-withdrawing groups, again due to their polar effects (Figure 3.13).

Other common groups that are electron withdrawing include trifluoromethyl (also trichloromethyl) and positively charged functional groups such as quarternary ammonium. In each case, these groups act mainly by polar effects. A trifluoromethyl group has three electronegative fluorine atoms attached to a single carbon rendering that carbon electron deficient. The nitrogen of a quaternary ammonium group has a formal positive charge that will attract electrons by electrostatics (Figure 3.14).

Figure 3.13 Overlap between a filled hybrid orbital (lone pair) on fluorine with an adjacent vacant σ*-orbital efficiently pushes some electron density back onto carbon, partially counteracting the strong polar effect and making fluorine a relatively weak electron-withdrawing group.

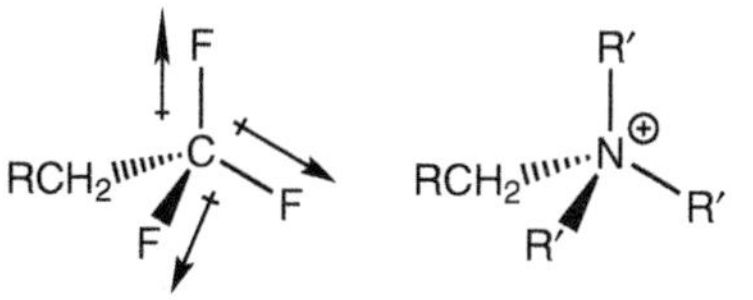

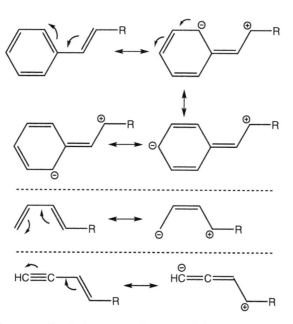

Figure 3.14 Left: A trifluoromethyl group is a strong electron-withdrawing group due to the combined polar effects of the three fluorine atoms. Right: A quaternary ammonium group is also a strong electron-withdrawing group due to the polar effect of the permanent positive charge on nitrogen.

Unsaturated groups such as phenyl, vinyl, ethynyl, etc. can also function as electron-withdrawing groups whenever they are attached to a system having an excess of electron density. In these cases, resonance is the main effect, with any charges being delocalized through the unsaturated system. More highly conjugated unsaturated systems will be more electron withdrawing than isolated double or triple bonds because there will be more possible resonance structures (Figure 3.15).

Figure 3.15 Unsaturated substituents as electron-withdrawing groups. Top: Aromatic rings allow for delocalization of electron density onto the *para*- and both *ortho*-positions of the ring. Middle: When an alkene is placed in conjugation with a source of electrons, they can be delocalized onto the more distant carbon of the double bond. Bottom: Alkynes also allow for delocalization of excess electron density.

There are far fewer functional groups that are *electron donating*. In most cases, these will include a heteroatom such as O, N, S, or P with one or more lone pairs of electrons and will be attached to an electron-deficient system by a σ-bond. Oxygen and nitrogen are more electronegative than carbon and will withdraw electrons by their polar effects, but donation by resonance will generally be the more dominant factor whenever possible. Sulfur has about the same electronegativity, and phosphorus is actually a little less electronegative as compared to carbon. Thus, these elements will not withdraw electrons by polar effects and instead can donate their lone pairs by resonance, although once again, the size difference between them and carbon makes such donation relatively inefficient. Typical examples of electron-donating functional groups include alcohols, ethers, thioethers, and amines (Figure 3.16).

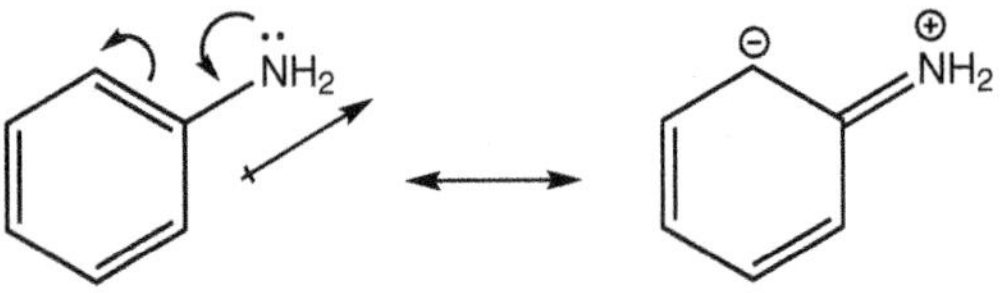

Figure 3.16 Amines withdraw electrons by their polar effects but donate by resonance. In general, resonance effects dominate whenever possible.

An interesting example that highlights the conflict between polar and resonance effects is seen with the three possible isomeric methoxyphenols as shown in Figure 3.17. In aqueous solution, phenols exist in equilibrium with their corresponding phenolate anion and a proton. The equilibrium constant, K, for *o*-methoxyphenol is 1.6 times larger than that of *p*-methoxyphenol. In both isomers, a lone pair of electrons from the methoxy group can be efficiently donated by resonance into the phenolate anion, which is destabilizing and is expected to shift the equilibrium toward the undissociated phenol. However, in *o*-methoxyphenol, the polar effect of the methoxy oxygen is strong and withdraws electron density from the phenolate anion and helps shift the equilibrium somewhat back toward the anionic form. The *para*-isomer however is too far away from the anionic charge for polar effects to be effective and so it mainly behaves as an electron-donor group. In *m*-methoxyphenol, resonance between the methoxy oxygen and the phenolate ion is not possible, yet the methoxy group is still close enough to the anionic charge for polar effects to withdraw electron density and stabilize the developing charge. Thus, a methoxy group at the *meta*-position behaves as an electron-withdrawing group and the equilibrium constant for this isomer is approximately twice that of *o*-methoxyphenol.

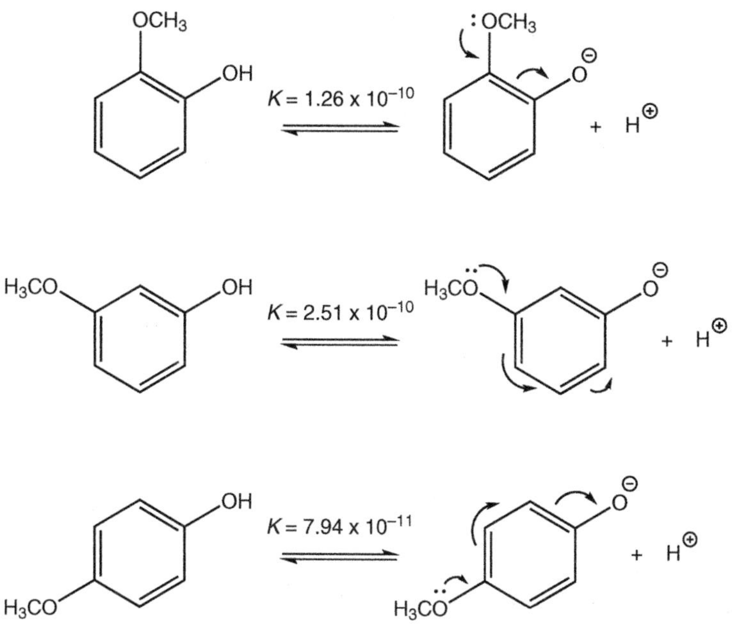

Figure 3.17 The dissociation of the three isomeric methoxyphenols. Top: An *ortho*-methoxy group donates electrons by resonance and withdraws by polar effects. Overall methoxy at the *ortho* position is a weak electron-donor group. Middle: The *m*-methoxy group cannot donate by resonance but is close enough to the phenol group for polar effects to be significant and so behaves as an electron-withdrawing group. Bottom: Methoxy at the *p*-position is a better electron-donating group because resonance is possible but polar effects are insignificant due to the distance of the group from the phenol. Equilibrium constants were obtained from reference [2].

Some heteroatom–containing functional groups that possess double bonds can be electron donating if the heteroatom rather than the double–bonded carbon is attached to the electron–deficient system by a single bond. Esters (shown in Figure 3.18), amides, carbonates, carbamates, amidines, and guanidines are a few of the functional groups for which this is true. It should be

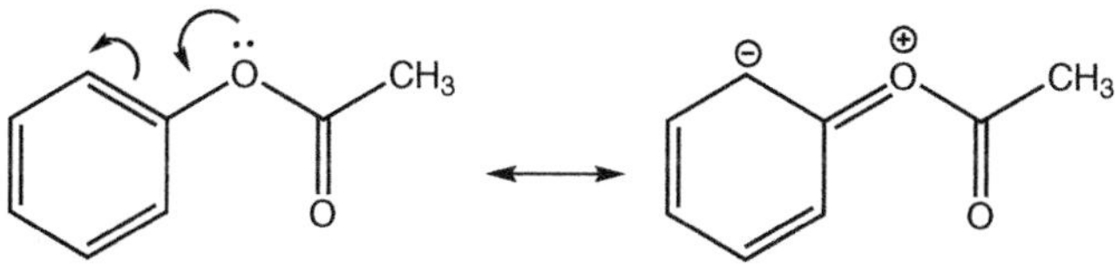

Figure 3.18 Attachment of an ester through the oxygen rather than the carbonyl carbon results in a group that is a weak electron donor by resonance.

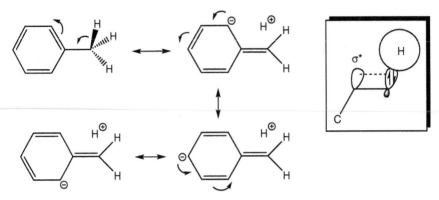

Figure 3.19 Hyperconjugation allows methyl groups to behave as electron donors. In the valence-bond explanation, electrons from a C–H bond are donated leaving a positive charge on the hydrogen. The shadow box shows the molecular orbital explanation of hyperconjugation which involves overlap of a filled C–H σ-orbital with a vacant σ*-orbital.

noted that because the heteroatom is attached to a carbonyl or C=N bond that will withdraw electron density from it, these groups tend to act as weaker electron donors than ethers or amines.

Simple alkyl groups (methyl, ethyl, propyl, etc.) also behave as weak electron-donating groups although it is not at first apparent how they can accomplish this since they lack a lone pair of electrons. The electrons that get donated come instead from the C–H bonds of the alkyl group in a process known as *hyperconjugation*. A valence-bond explanation for this phenomenon has resonance structures in which electrons from an adjacent C–H bond are pushed toward the site that is electron deficient putting a positive charge on the hydrogen, which is temporarily no longer bonded to carbon (Figure 3.19). The molecular orbital explanation for hyperconjugation involves overlap of a filled C–H σ-bond with a vacant adjacent C–C σ*-orbital. Methyl groups are the most strongly donating of the alkyl groups because they have the greatest number of C–H bonds (three) followed by methylene (CH_2) and methine (CH) groups.

It was noted previously that unsaturated functional groups such as phenyl, alkenyl, and alkynyl can be electron withdrawing by delocalizing developing negative charges. Those same functional groups can also function as electron-donating groups in response to developing positive charges. Again it is resonance that allows for delocalizing the charge (Figure 3.20).

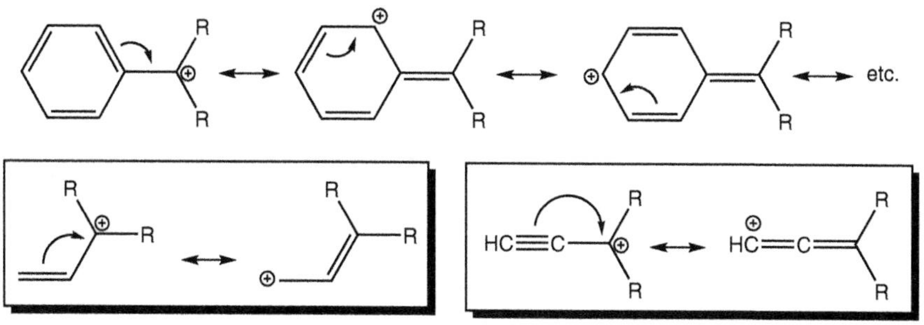

Figure 3.20 Aromatic rings, alkenes, and alkynes can also serve as electron-donating substituents.

REFERENCES

[1] E.J. Grubbs, R. Fitzgerald, Angular dependence of substituent effects in geometrically isomeric 11,12-dichloro-9,10-ethanoanthracene, Tetrahedron Letters 47 (1968) 4901–4904.

[2] Z.B. Alfassi, R.H. Schuler, Reaction of azide radicals with aromatic compounds. Azide as a selective oxidant, Journal of Physical Chemistry 89 (1985) 3359–3363.

CHAPTER 4

Acids and Bases*

Drug molecules find themselves in an aqueous environment once they enter the body. Depending on the nature of whatever functional groups the drug possesses, it will generally behave as either an acid or a base and will exist to a certain extent as a charged species. The degree to which this occurs will play a role in determining how that drug gets distributed throughout the body. Thus an understanding of the concepts of acids and bases is a critical aspect of medicinal chemistry. The definition of the terms acid and base however depends upon which of the two theories (*Brønsted–Lowry* or *Lewis*) is most applicable to the situation at hand. When dealing with aqueous solutions, the Brønsted–Lowry theory is most commonly invoked, although Lewis acidity and basicity is still important. In this chapter, we will focus first on the Brønsted–Lowry theory in which an *acid* is defined as a group that can donate a proton (a hydrogen cation) and a *base* is a group that can accept a proton. Lewis's theory is much broader and states that an acid is a group that can accept a pair of electrons while a base is a group that can donate a pair of electrons. This will be discussed in more detail in the final section of this chapter.

BRØNSTED–LOWRY ACIDS AND BASES

Acidic functional groups are those that contain weakly bound hydrogen that can be easily removed by a base. For the most part this is limited to O–H, N–H, S–H, and P–H bonds with the electronic effects of the functional groups attached to the O, N, S, and P helping to determine whether the hydrogen can in fact be donated (behave as an acid). Generally most C–H bonds are too strong to be removed in aqueous solution, although there are exceptions that will be mentioned later in this chapter. Any attempt to employ a very strong base to deprotonate a carbon will result instead in a proton being lost from water. Thus the strongest base that can exist in water is hydroxide anion, OH^-. To remove C–H bonds using strong

* "Material in this chapter on pages 70, 72 and 73 and Tables 4.1, 4.2 and 4.4 first appeared in Remington's Pharmaceutical Sciences, 19th edition and is used with permission of Philadelphia College of Pharmacy."

Organic Chemistry Concepts and Applications for Medicinal Chemistry
http://dx.doi.org/10.1016/B978-0-12-800739-6.00004-8

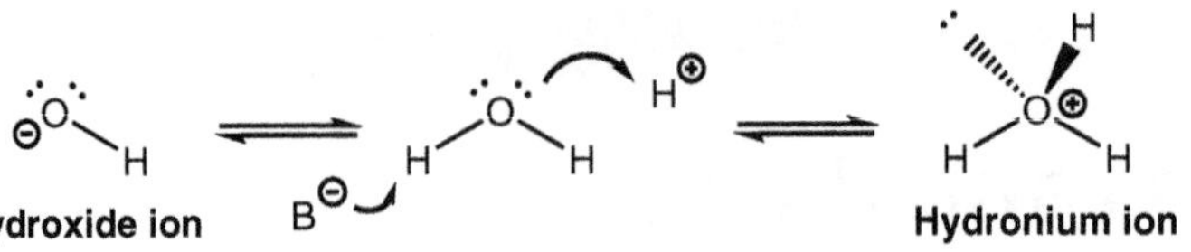

Figure 4.1 Water can function as an acid or a base making it an amphoteric substance. As a base, water accepts protons to form the hydronium ion. A base can remove one of the water's protons to form hydroxide ion. These two species are, respectively, the strongest acid and base that can exist in water.

bases requires the use of solvents that are much less acidic than water such as hydrocarbons or ethers. While this is extremely important for drug synthesis, it is beyond the scope of this book (Figure 4.1).

When an acid is placed into an aqueous solution the proton is donated to any basic species present in that solution (including water itself). Once that proton has been donated, a new species is formed which has a charge one less than the acid itself. This species is called the *conjugate base* of the acid. If the acid is electrically neutral (uncharged), then the conjugate base has a charge of -1. Acids that are positively charged have conjugate bases that are electrically neutral. The acid and its conjugate base are in *equilibrium*, and thus there is always a certain concentration of each species present in solution. The relative amount of each species is determined by the size of the equilibrium constant, K_a, called the *acid dissociation constant* (Figure 4.2).

The magnitude of K_a is determined by the relative stabilities of the acid and its conjugate base. Electron-withdrawing groups will stabilize negatively

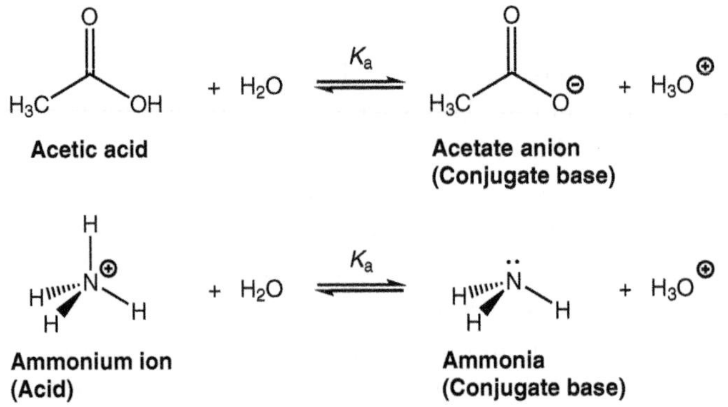

Figure 4.2 An electrically neutral acid such as acetic acid will dissociate in water to form a negatively charged conjugate base. Any acid that is charged, such as ammonium ion, will dissociate to form a conjugate base with a charge one unit less than the acid.

charged conjugate bases relative to the acids from which they are derived. This serves to increase the magnitude of K_a meaning that such acids will be stronger than if they were lacking a withdrawing group. Positively charged acids will also be stronger if they have electron-withdrawing substituents. These would destabilize a positive charge and therefore shift the equilibrium toward the conjugate base (uncharged). Electron-donating groups have the opposite effect. They destabilize negatively charged conjugate bases, shifting the equilibrium back toward the uncharged acid. With charged acids, an electron-donating group stabilizes the positive charge again driving the equilibrium toward the acid. Thus, acids substituted with electron-donating groups are weaker than acids lacking such substituents (Figure 4.3).

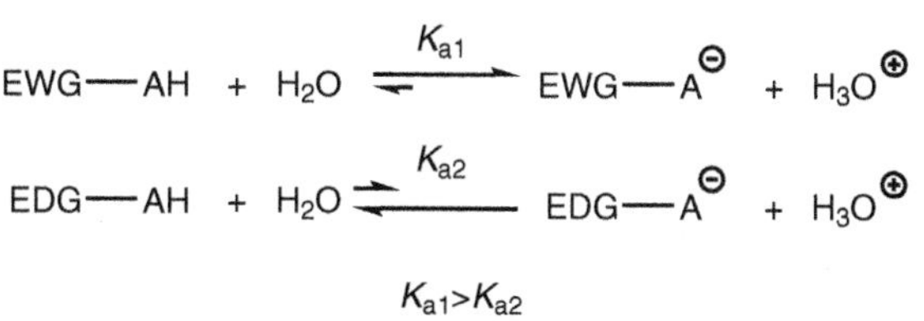

$$EWG-AH \ + \ H_2O \ \underset{}{\overset{K_{a1}}{\rightleftharpoons}} \ EWG-A^{\ominus} \ + \ H_3O^{\oplus}$$

$$EDG-AH \ + \ H_2O \ \underset{}{\overset{K_{a2}}{\rightleftharpoons}} \ EDG-A^{\ominus} \ + \ H_3O^{\oplus}$$

$$K_{a1} > K_{a2}$$

Figure 4.3 An electron-withdrawing group (EWG) attached to an acid will stabilize the conjugate base of the acid and increase K_a. An electron-donating group (EDG) will destabilize the conjugate base to a greater extent than the acid and will decrease K_a.

Bases must have a donatable pair of electrons to allow for formation of a bond with a proton. Typically many nitrogen-containing functional groups (amines, imines, guanidines, and amidines) behave as bases because the lone pair electrons on the nitrogen can be donated. Oxygen-containing functional groups (alcohols, ethers, etc.) are generally not basic, or only very weakly basic, because the higher electronegativity of oxygen as compared to nitrogen inhibits sharing of the lone pair electrons. The exception is when oxygen carries a negative charge as in alkoxide and hydroxide ions.

When a base forms a bond with a proton, its charge will increase by one unit and the resulting species is called the *conjugate acid* of the base. Conjugate acids of electrically neutral bases have a charge of +1 while those derived from negatively charged bases are electrically neutral (Figure 4.4).

The reaction of a base with a proton to form a conjugate acid is reversible and equilibrium is established between these components. For a base, the equilibrium constant is called the *base association constant*, K_b. When an equation is written in which the acid component is to the left of the equilibrium arrows, K_a is the equilibrium constant, whereas when the basic component is left of the equilibrium arrows, it is proper to use K_b for the

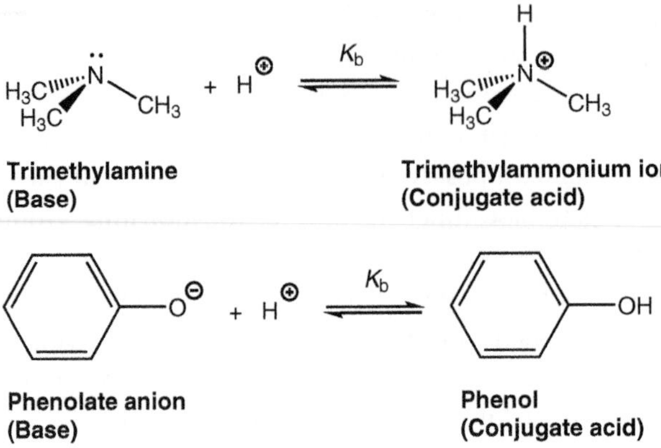

Trimethylamine
(Base)

Trimethylammonium ion
(Conjugate acid)

Phenolate anion
(Base)

Phenol
(Conjugate acid)

Figure 4.4 An electrically neutral base will become protonated to form a conjugate acid with a charge of +1. An anionic base will react with a proton to form a conjugate acid that is not charged.

constant. It is a common practice to take the negative of the common logarithm of the equilibrium constant $(-\log K_a)$ and refer to it as the pK_a to avoid having to write exponential expressions. Thus if $K_a = 1 \times 10^{-5}$, then the pK_a would simply be 5. A weaker acid with, for example, $K_a = 1 \times 10^{-10}$ would have a pK_a of 10. It is seen therefore that stronger acids have low pK_a values while for weaker acids the pK_a will be higher. Base association constants can be treated similarly, thus $pK_b = -\log K_b$. A strong base with $K_b = 1 \times 10^{-2}$ has a pK_b of 2; a weak base with $K_b = 1 \times 10^{-8}$ has $pK_b = 8$. Therefore, strong bases have low pK_b values and weak bases have high pK_b values. In water, both the pK_a and pK_b scales extend from 0 to 14. There is a relationship that pertains to acids and bases dissolved in water:

$$pK_a \, (\text{acid}) + pK_b \, (\text{conjugate base}) = 14$$

$$pK_b \, (\text{base}) + pK_a \, (\text{conjugate acid}) = 14$$

Thus if the pK_a of an acid is known, the pK_b of its conjugate base can be easily calculated. Similarly, if the pK_b of a base is known, then the pK_a of its conjugate acid can be determined. For example, looking at the structures in Figure 4.5, if you know that acetic acid has a $pK_a = 4$, then you can calculate that acetate anion (its conjugate base) has a $pK_b = 10$; and if you know that trimethylamine has a $pK_b = 5$, then trimethylammonium ion (its conjugate acid) has a $pK_a = 9$.

In many books, both acid and base strength are reported as pK_a values. It must be understood that when referring to bases using pK_a values, it is the

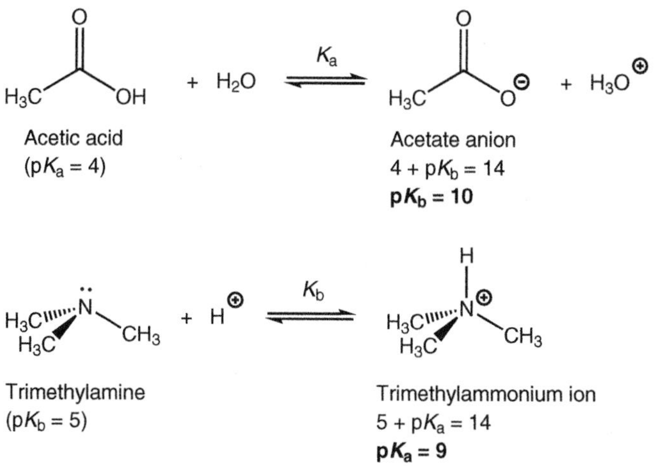

Figure 4.5 If the pK_a of an acid is known, then the pK_b of its conjugate base is easily calculated. Using the same relationship, if the pK_b of a base is known, then the pK_a of its conjugate acid may be determined.

strength of the conjugate acid of the base that is actually being reported. According to the relationships defined above, a strong base would have a large pK_a value for its conjugate acid, and a weak base would have a low value for the pK_a of its conjugate acid. Thus, weak bases have low pK_a values and strong bases have high pK_a values. In this chapter, base strength will be referred to using pK_b values to avoid confusion.

When an acid reacts with a base, the product is called a *salt*. Salts are ionic compounds with a cationic portion that derives from the base and a negatively charged counterion that originates from the acid. As ionic compounds, they are generally highly water soluble, a property that accounts for the fact that most drugs are administered in salt form (Figure 4.6).

IONIZATION

In an acid–base equilibrium, at least one species always will have a charge (will exist in an *ionized* state in solution). The relative amount of the charged and uncharged species plays an important part in determining the biological activity of drug molecules. In the body, a drug must be able to pass from one compartment to another by crossing through lipid bilayers. These bilayers have both polar and nonpolar regions and molecules that are extensively ionized might have difficulty passing through the nonpolar regions of bilayers, while other molecules that are largely uncharged may have

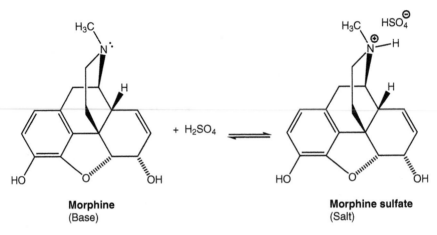

Morphine
(Base)

Morphine sulfate
(Salt)

Figure 4.6 Most drugs are administered as salts because in that form they are largely water soluble. Here, we see how morphine, which has a basic amino group, reacts with sulfuric acid to form the salt morphine sulfate. This is classified as a 1:1 salt because it is made up of 1 mol of base and 1 mol of acid.

trouble traveling through the polar domains. It is also important to realize that different parts of the body have different pH ranges (e.g. the stomach is at pH 1–3 while blood plasma is pH 7.4) and it therefore becomes important to understand the effect of pH on the degree of ionization of a drug molecule.

Let HA represent any uncharged acid with A^- as its conjugate base.

$$HA \xrightleftharpoons{K_a} A^{\ominus} + H^{\oplus} \qquad [4.1]$$

By definition: $K_a = [A^-][H^+]/[HA]$

Take the logarithm of each side of the equation:

$$\log K_a = \log [A^-] + \log [H^+] - \log [HA]$$

Now multiply each side of the equation by -1:

$$-\log K_a = -\log [A^-] - \log [H^+] + \log [HA]$$

Substitute pK_a for $-\log K_a$ and pH for $-\log [H^+]$:

$$pK_a = pH + \log [HA] - \log [A^-]$$

Now combine the two log terms:

$$pK_a = pH + \log [HA]/[A^-]$$

This is known as the *Henderson–Hasselbalch* equation and it relates the amount of an acid of known pK_a to the extent to which it is ionized at any given pH. It is important to understand that pK_a is a property of the molecule (it depends on the chemical structure) whereas pH is a property of the medium (solvent) into which that compound is placed. Below is an example of using the Henderson–Hasselbalch equation to determine the degree to which acetic acid is ionized at pH 7.

$$CH_3CO_2H \rightleftharpoons CH_3CO_2^{\ominus} + H^{\oplus} \qquad [4.2]$$

$$pK_a = 4 \qquad pH = 7$$

Both the pH (7) and pK_a (4) of acetic acid are known and the values can be plugged into the above equation to give

$$4 = 7 + \log [CH_3CO_2H]/[CH_3CO_2^-]$$

Subtraction gives

$$-3 = \log [CH_3CO_2H]/[CH_3CO_2^-]$$

Take the antilog of both sides of the equation to get

$$1/1000 = [CH_3CO_2H]/[CH_3CO_2^-]$$

Thus, the ratio of unionized acetic acid to acetate anion (ionized) at pH 7 is 1:1000.

Another way to view this is to express the result as % *ionization*.

$$\% \text{ ionization} = 100(\text{ionized})/[(\text{ionized}) + (\text{unionized})]$$

In the example above, acetate anion is the ionized species and acetic acid is unionized. Use the amounts from the ratio obtained from the Henderson–Hasselbalch equation to plug into this new equation:

$$\% \text{ ionization} = 100(1000)/(1000 + 1)$$

$$\% \text{ ionization} = 99.9\%$$

In other words at pH 7, acetic acid exists mostly as the ionized species, acetate anion.

Let us now examine the behavior of a charged acid, such as ammonium ion. Let BH^+ represent any charged acid, with B as its conjugate base.

$$BH^{\oplus} \underset{}{\overset{K_a}{\rightleftharpoons}} B: + H^{\oplus} \qquad [4.3]$$

By definition $K_a = [B][H^+]/[BH^+]$

Take logarithms of both sides of the equation:

$$\log K_a = \log [B] + \log [H^+] - \log [BH^+]$$

Multiply both sides of the equation by -1:

$$-\log K_a = -\log [B] - \log [H^+] + \log [BH^+]$$

Substitute pK_a for $-\log K_a$ and pH for $-\log[H^+]$:

$$pK_a = pH + \log [BH^+]/[B]$$

This is the Henderson–Hasselbalch equation for charged acids. Note that in this case in the logarithmic term, it is the ratio of the ionized form divided by the unionized form. But here the ionized form is the acid and the unionized form is the conjugate base. Thus the Henderson–Hasselbalch equation can be generalized as

$$pK_a = pH + \log [\text{acid form}]/[\text{base form}]$$

Remember: the base goes in the basement.

For uncharged acids, solution of this equation gives the ratio of unionized/ionized; for charged acids, solution of the equation gives the ratio of ionized/unionized.

ESTIMATING ACID/BASE STRENGTH

Conjugate bases are more electron rich than the acids from which they derive. As mentioned earlier, electron-withdrawing groups stabilize these species and drive the equilibrium to the right, thereby increasing the strength of the acids. Table 4.1 lists a series of structures along with their

Table 4.1 The Effect on pK_a of Substitution of One of Water's Hydrogen Atoms by Various Substituents

Substituent	Structure	pK_a
Hydrogen	H–OH	14
Alkyl	R–OH	14
Alkene	$R_2C{=}CR{-}OH$	9
Aryl	Ar–OH	9
Acyl (carbonyl)	$R(C{=}O){-}OH$	4
Sulfonyl	$RSO_2{-}OH$	<1

corresponding average pK_a values. Each structure can be viewed as being derived from water by replacement of one hydrogen atom by a different group.

Examination of the data in this table reveals several interesting facts. First, there is no difference between the pK_a values for H–OH (water) and R–OH (alcohol). Both have pK_a values of 14 and therefore behave as extremely weak acids. If one of the hydrogen atoms of water is replaced instead with an electron-withdrawing group such as a C=C bond or an aromatic ring, the resulting compounds (enols and phenols, respectively) are more acidic than water by $5\,pK$ units. Replacement of hydrogen with an even stronger electron-withdrawing group, such as a carbonyl, results in formation of a carboxylic acid which is more acidic than water by $10\,pK$ units. Finally, replacement of hydrogen with a sulfonyl moiety gives rise to a sulfonic acid which is a very strong acid, with a pK_a of <1. Thus we note that stronger electron-withdrawing groups do in fact increase acidity. If both of the hydrogens of water are replaced by other functional groups, the resulting compounds will no longer be acidic since there will be no donatable hydrogen. Such compounds are said to be *neutral*.

Now we will repeat this exercise using the same set of functional groups, but this time attached to a NH_2 group (Table 4.2). Base strength depends on the ease with which that pair of electrons can be donated. We would therefore expect that electron-withdrawing groups would decrease base strength. We see that once again whether the NH_2 group is attached to a hydrogen (ammonia) or an alkyl group (primary amine) the base strength is identical, but attaching either an alkene (enamine) or aryl (arylamine) results in an increase of their pK_b values by 5 units relative to ammonia, meaning that these are indeed weaker bases. These groups can delocalize some of the electron density from the nitrogen onto the double bond or aromatic ring. It is more difficult to donate a charge that is delocalized than one which is concentrated on a

Table 4.2 The Effect on pK Value of Substitution of One of Ammonia's Hydrogen Atoms by Various Groups

Substituent	Structure	pK Value
Hydrogen	$H-NH_2$	$pK_b = 5$ (base)
Alkyl	$R-NH_2$	$pK_b = 5$ (base)
Alkene	$R_2C=CR-NH_2$	$pK_b = 10$ (base)
Aryl	$Ar-NH_2$	$pK_b = 10$ (base)
Acyl (carbonyl)	$R(C=O)-NH_2$	$pK_a = 14$ (acid)
Sulfonyl	RSO_2-NH_2	$pK_a = 9$ (acid)

single atom. As we move down in Table 4.2, we notice that replacement of hydrogen by a carbonyl group gives rise to an amide that has a pK_a value rather than a pK_b value. This implies that amides behave as acids (albeit weak ones) and that the nitrogen is no longer basic. A carbonyl group is a strong electron-withdrawing group that exerts both a polar and resonance effect on the adjacent nitrogen rendering the lone pair incapable of being donated. Thus amides are not bases (at least according to the Brønsted–Lowry definition). As long as the amide nitrogen has at least one hydrogen atom attached, it will behave as an acid. Tertiary amides that have two R groups attached to nitrogen cannot be acids and since they are also not bases, they are neutral. A sulfonyl group is an even stronger electron-withdrawing group than carbonyl and when such a group is attached to nitrogen to form a sulfonamide, the nitrogen is again no longer basic. When a proton dissociates from nitrogen, the resulting conjugate base form is stabilized by polar and resonance effects from the sulfonyl group and the acidity is equivalent to that of a phenol. If a sulfonamide lacks hydrogen on the nitrogen, then it is neutral (Figure 4.7).

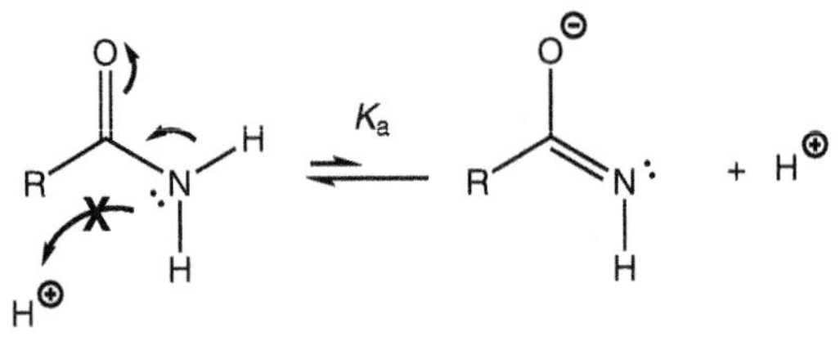

Figure 4.7 In amides, the electron-withdrawing effect of the carbonyl exerts such a strong attraction for the lone pair of electrons on the nitrogen that it is no longer able to form a bond with a proton, and is therefore no longer considered to be a base. Instead, if the nitrogen has an attached hydrogen, then that hydrogen can dissociate. This means that amides behave as weak acids.

In Table 4.3, we see the effect of replacing the second hydrogen on nitrogen by various functional groups. Not surprisingly, a second alkyl group does not significantly change the pK_b value relative to a primary alkylamine. Similarly, an alkyl group does not alter the basicity of an arylamine or the acidity of an amide or sulfonamide. An aryl group when attached to the nitrogen of an arylamine, amide, or sulfonamide does however have an effect on the pK value. The pK_b of an arylamine increases to 12 when a second aryl group is added. When aryl is attached to the nitrogen of an amide, the pK_a decreases by 2 units to 12. In a similar fashion, an N-arylsulfonamide is 2 units more acidic ($pK_a = 7$) than a sulfonamide. Attachment of a second carbonyl group to the nitrogen of an amide gives rise to an imide ($pK_a = 9$), that is, 5 units

Table 4.3 The Effect on pK Value of Substituting for Two of Ammonia's Hydrogen Atoms

Functional Group	Structure	pK Value
2° Alkylamine	R–NH–R	pK_b = 5 (base)
2° Arylamine	Ar–NH–R	pK_b = 10 (base)
Diarylamine	Ar–NH–Ar	pK_b = 12 (base)
N-Alkylamide	R(C=O)–NH–R	pK_a = 14 (acid)
N-Arylamide	R(C=O)–NH–Ar	pK_a = 12 (acid)
Imide	R(C=O)–NH–(C=O)R	pK_a = 9 (acid)
N-Alkylsulfonamide	RSO$_2$–NH–R	pK_a = 9 (acid)
N-Arylsulfonamide	RSO$_2$–NH–Ar	pK_a = 7 (acid)
N-Acylsulfonamide	RSO$_2$–NH–(C=O)R	pK_a = 4 (acid)
N-(Sulfonyl)sulfonamide	RSO$_2$–NH–SO$_2$R	pK_a = 2 (acid)

more acidic than an amide. A sulfonamide with a carbonyl attached to the nitrogen is called an N-acylsulfonamide. It too is 5 units more acidic than a sulfonamide with a pK_a value of 4. Finally, an N-(sulfonyl)sulfonamide is more acidic than a sulfonamide by 7 units (pK_a = 2).

Tables 4.1 and 4.2 show the effect on pK_a or pK_b of replacing one hydrogen atom of either water or ammonia by various functional groups. In Table 4.3 the effect of replacing two hydrogen atoms of ammonia by those same functional groups is given. These results are summarized and generalized in Table 4.4. It is important to realize that these values are ball park approximations. For example, not all phenols will have the same pK_a value and different alkylamines will vary in basicity from each other and from ammonia. The question remains, however, how these values can be used to predict pK_a and pK_b values for a wide range of functional groups. To do that requires that one consider the pK_a and pK_b scales to be a continuum. Arrangement of the scales so that pK_b increases from 0 to 14 on the left and pK_a decreases from 14 to 0 on the right gives rise to the pK_a and pK_b calculator shown in Figure 4.8. To use the calculator, decide whether the functional group in question can be

Table 4.4 The Generalized Effect on pK Value of Various Functional Groups as a First Substitution on Water or a First or Second Substitution on Ammonia

Substituent	First-Group Effect (pK units)	Second-Group Effect (pK units)
H, alkyl	0	0
Aryl, alkene, alkyne	5	2
Carbonyl	10	5
Sulfonyl	15	7

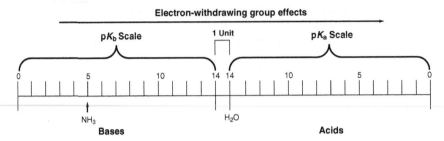

Figure 4.8 The pK_a–pK_b calculator. Note that attachment of electron-withdrawing groups will increase pK_b values and decrease pK_a values.

made by substituting for hydrogen of water or ammonia and use that location on the calculator as a starting point. Then make note of which substituent(s) are directly connected to the oxygen or nitrogen. For a single substituent, use the first-group effect from Table 4.4 and move that number of units to the right along the calculator (counting the gap between the pK_b and pK_a scales as 1 unit). For two substituents that are attached to nitrogen, use ammonia (pK_b = 5) as the starting point. Select the substituent that has the larger effect as the first group and for the other one, use the second-group effect. Again, you always move to the right along the calculator. For example, if nitrogen was substituted with a sulfonyl group and a carbonyl group, you would select the sulfonyl for the first group and move 15 units to the right and then since carbonyl was the second group, move an additional 5 units to the right giving a final pK_a of 4. It is important however that you apply chemical sense to the process. If the sulfonamide that was just calculated to have a pK_a of 4 lacks hydrogen on the nitrogen, then it cannot have a pK_a value (since there can be no equilibrium) and it must be neutral.

The reader might at this point be wondering why there is no third group effect listed in Table 4.4 since ammonia has three hydrogen atoms. Attachment of a third group to nitrogen introduces considerable crowding around the lone pair (steric effects). While the third substituent will still exert an electronic effect, steric effects may make it difficult for the nitrogen to donate the lone pair electrons. These effects are much more difficult to predict as compared to having only one or two groups on the nitrogen. In such cases, estimates cannot be made by simple inspection. Instead one must rely on chemical intuition to make predictions of properties. For example, one can calculate that diphenylamine has a pK_b value of 12 using the guidelines outlined above. If a third phenyl group was attached to nitrogen to make triphenylamine, it would be reasonable to assume that the extra

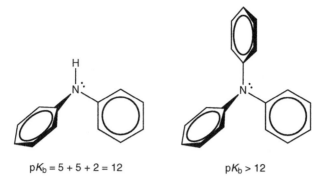

$pK_b = 5 + 5 + 2 = 12$ $pK_b > 12$

Figure 4.9 The pK_b value for diphenylamine can be readily estimated using the calcula-tor shown in Figure 4.8. The attachment of a third phenyl, however, introduces steric factors that make predicting actual pK_b values difficult. Thus we estimate that triphenyl-amine will still be basic, but less basic than diphenylamine (p$K_b > 12$).

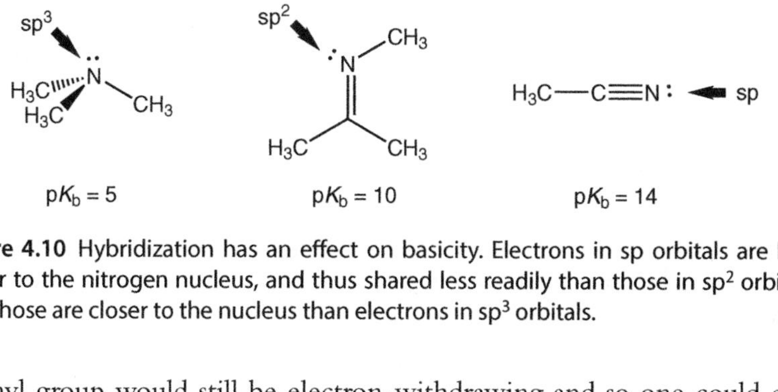

$$sp^3 \qquad sp^2$$

$pK_b = 5$ $pK_b = 10$ $pK_b = 14$

Figure 4.10 Hybridization has an effect on basicity. Electrons in sp orbitals are held closer to the nitrogen nucleus, and thus shared less readily than those in sp^2 orbitals, and those are closer to the nucleus than electrons in sp^3 orbitals.

phenyl group would still be electron-withdrawing and so one could esti-mate the pK_b of triphenylamine to be greater than 12 (Figure 4.9).

In addition to substitution by electron-withdrawing groups, another fac-tor that helps to determine the basicity of nitrogen is the hybridization of the orbital containing the lone pair. Ammonia and alkylamines are moder-ately-strong bases with pK_b values of 5. The lone pair is housed within an sp^3 hybrid orbital. Imines have their lone pairs in sp^2 hybrid orbitals while in a nitrile, the lone pair resides in an sp hybrid orbital. Recall from Chapter 1 that as hybridization changes from sp^3 to sp^2 to sp, the corresponding bond lengths tend to decrease and electronegativity increases. This implies that imines should be weaker bases than amines and nitriles should be weaker bases than imines. As a rough approximation, the average pK_b value for imines is 10 while for nitriles it is 14 (Figure 4.10). Aromatic heterocycles such as pyridine, pyrimidine, thiazole, and so on, all have sp^2-hybridized

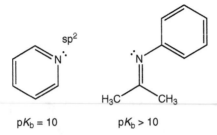

pKb = 10   pKb > 10

Figure 4.11 Left: Certain unsaturated heterocycles, such as pyridine, also have the lone pair in a sp^2-hybrid orbital and thus have pK_b values equivalent to those of imines. Right: Attachment of electron-withdrawing groups will raise the pK_b of imines, but to an unpredictable degree. It is proper to estimate the pK_b as being >10.

nitrogens within their structures and these nitrogens will also have pK_b values of approximately 10. It is reasonable to assume that the attachment of electron-withdrawing groups to an imine would decrease its base strength although the extent to which that occurs is not as easily predicted as for sp^3-hybridized nitrogen. Thus an N-phenylimine, for example, would have a pK_b value greater than 10 (Figure 4.11).

A different situation arises if an electron-donating group such as a saturated nitrogen with a lone pair is attached to the sp^2 carbon of an imine. The lone pair electrons from the saturated nitrogen can be donated by resonance to the sp^2 nitrogen, increasing the electron density at this position, and thereby making the sp^2 nitrogen a much stronger base than an imine. These electrons can also stabilize the conjugate acid by delocalizing the charge—an effect that serves to increase the base strength, much as stabilization of conjugate bases increased the strength of acids. Such a functional group is called an amidine, and the sp^2 nitrogen of the amidine has $pK_b = 3$. The sp^3 nitrogen of an amidine is an extremely weak base with a pK_b value near 14. This is due to the lone pair electrons being delocalized into the $C=N$ bond. If an electron-withdrawing group is attached to either nitrogen of an amidine, its basicity will be decreased and the pK_b value of the sp^2-hybridized nitrogen will be >3 (Figure 4.12).

Guanidines are similar to amidines except that they have an additional saturated nitrogen that can donate electrons into the $C=N$ bond. The sp^2 nitrogen of a guanidine is a very strong base with a pK_b value of 1. The saturated (sp^3) nitrogens of guanidines both have pK_b values of 14 (very weak bases). As with amidines, attachment of electron-withdrawing groups onto any of the nitrogens of a guanidine makes it a weaker base (Figure 4.13).

Toward the beginning of this chapter, it was mentioned that in aqueous solution $C–H$ bonds are usually not considered to be acidic. While this is

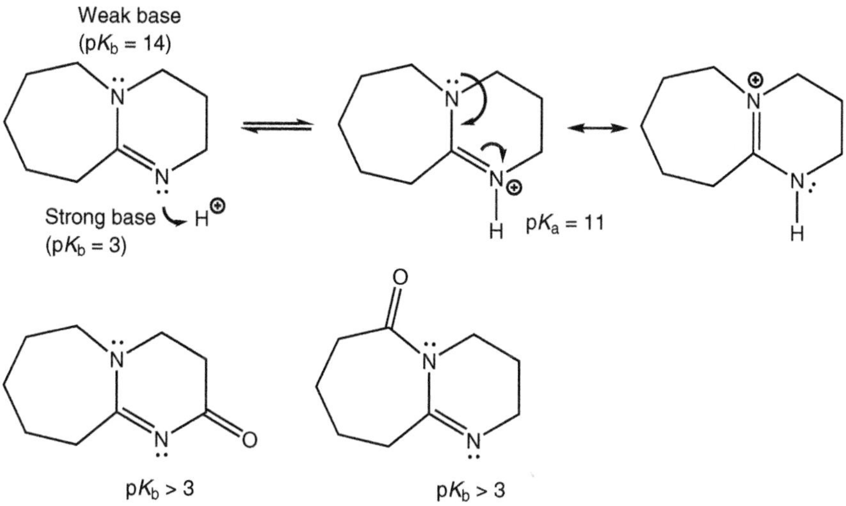

Figure 4.12 Top: A cyclic amidine. Note that the sp²-hybridized nitrogen is strongly basic, while the sp³ nitrogen is a very weak base. If the sp² nitrogen becomes protonated, then the charge can be delocalized by donation from the sp³ nitrogen. Bottom: Attachment of electron-withdrawing groups to either the sp² nitrogen or the sp³ nitrogen will lower the base strength of amidines.

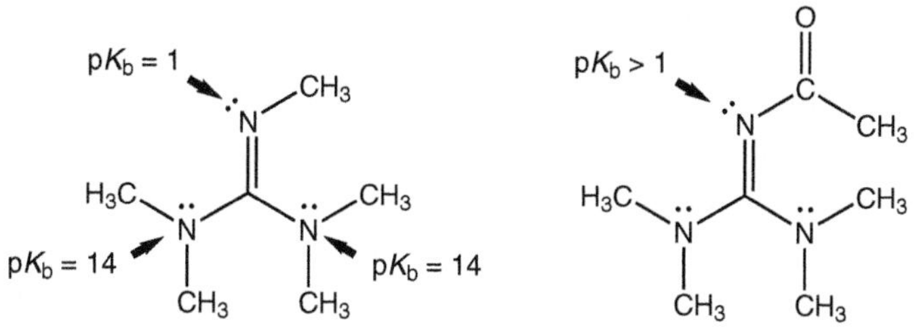

Figure 4.13 Left: The structure of pentamethylguanidine, showing again that it is the sp²-hybridized nitrogen that is strongly basic while the two sp³ nitrogens are only weakly basic. Right: As with amidines, electron-withdrawing groups will increase the pK_b of guanidines.

generally true, there are certain exceptions. Remember from Chapter 3 that carbonyl compounds often exist in equilibrium with their tautomeric enol forms. For simple ketones and esters, the enol form is very minor and does not contribute enough to have an impact on acidity. The compound acetylacetone (Figure 3.6), however, demonstrated that certain structural features can actually lead to the enol form being favored. Since enols are acidic ($pK_a = 9$) compounds with high enol content are considered to be carbon

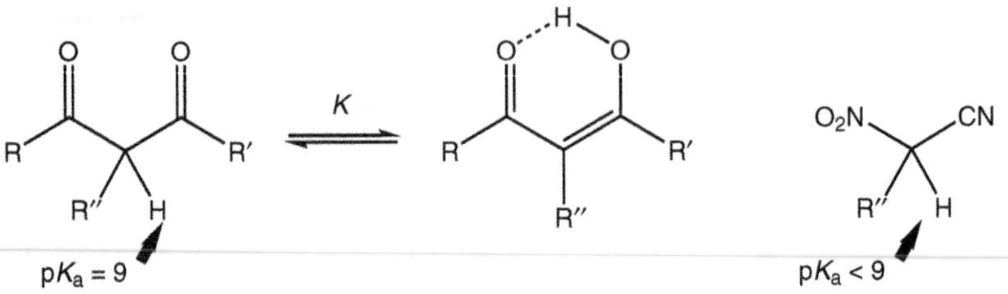

Figure 4.14 Left: A CH group that bears two carbonyl-based functional groups will have a significant amount of the enol tautomer. This in essence makes the CH group acidic with a pK_a value of 9. Right: Stronger electron-withdrawing groups attached to a CH group will greatly weaken the C–H bond making it an even stronger acid.

acids. Thus a C–H or CH_2 group that is attached to two carbonyl groups will be acidic with a pK_a value of approximately 9. Stronger electron-withdrawing groups such as nitro or nitrile will weaken the C–H bond between them even more resulting in more acidic compounds with pK_a values <9 (Figure 4.14).

LEWIS ACIDS AND BASES

In the beginning of this chapter, the Lewis definition of acids and bases was mentioned briefly. This last section will explore this concept in a little more detail. According to Lewis, an acid is a group that can accept a pair of electrons. This is quite different from the Brønsted–Lowry definition of an acid. To accept a pair of electrons requires an easily accessible vacant orbital. Elements of Group III of the Periodic Table tend to be sp^2-hybridized with trigonal planar geometry. Since these elements have only three valence electrons, the remaining p-orbital is vacant. Hence many boron and aluminum compounds (the first two members of this group) act as Lewis acids. Many of the main group elements in rows 3 and beyond have vacant d-orbitals, and therefore, compounds such as SO_3 and $SnCl_4$ also function as Lewis acids. The same is true for many of the transition metals such as zinc, iron, and titanium. Finally, any cationic species in which the charge originates from a vacant orbital (unlike ammonium ion) is a Lewis acid. A proton, by virtue of having a vacant 1s orbital, is also a Lewis acid (Figure 4.15).

Lewis bases are groups that can donate a pair of electrons. This definition is similar to the Brønsted–Lowry definition, but turns out to be broader in scope. Elements with lone pair electrons are all Lewis bases including ethers, carbonyl oxygens, amines, phosphines, and so on. Also anionic species are Lewis bases. Electrons in π-bonds also fall within this definition and

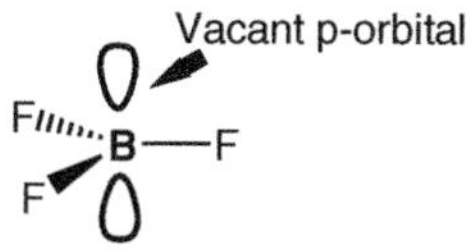

Vacant p-orbital

Figure 4.15 Boron trifluoride is a classic Lewis acid. The hybridization of boron is sp^2 and boron, therefore, also has a p-orbital that is vacant, and can accept a pair of electrons.

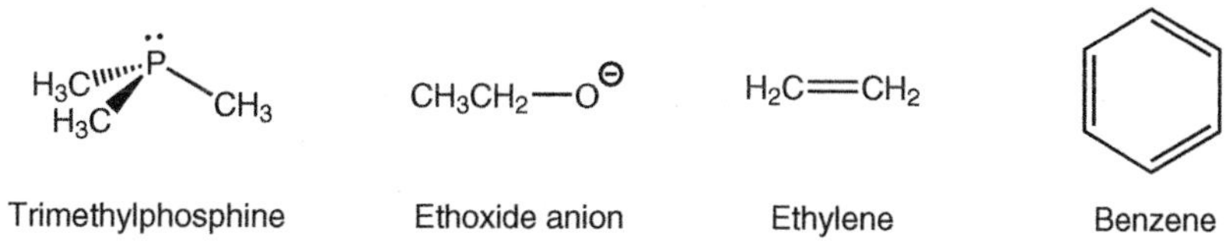

Trimethylphosphine	Ethoxide anion	Ethylene	Benzene

Figure 4.16 Several examples of Lewis bases are shown here. Compounds with lone pairs and formal negative charges would also fit the Brønsted–Lowry definition of a base. Here we see that electrons in π-bonds can also be shared and so alkenes, alkynes, and aromatic rings are considered to be Lewis bases.

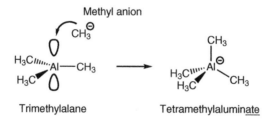

Methyl anion

Trimethylalane   Tetramethylaluminate

Figure 4.17 A Lewis acid (trimethylalane) reacts with a Lewis base (methyl anion) to form a complex in which the central atom of the acid (Al) now has a negative charge. Species such as this are known as "ate" complexes (here it is called tetramethylaluminate).

so according to Lewis, alkenes, alkynes, and aromatic rings are all bases (Figure 4.16).

The reaction of a Lewis acid with a Lewis base results in formation of a complex or salt, much as Brønsted acids and bases produce salts. When a Lewis acid reacts with a Lewis base and the acidic center acquires a negative charge as a result of having a higher than normal valence, the resulting salt is called an "*ate complex*" (Figure 4.17).

When it is the Lewis base that expands its normal valence resulting in the basic atom becoming positively charged, the resulting product is called an "*onium salt*" (Figure 4.18).

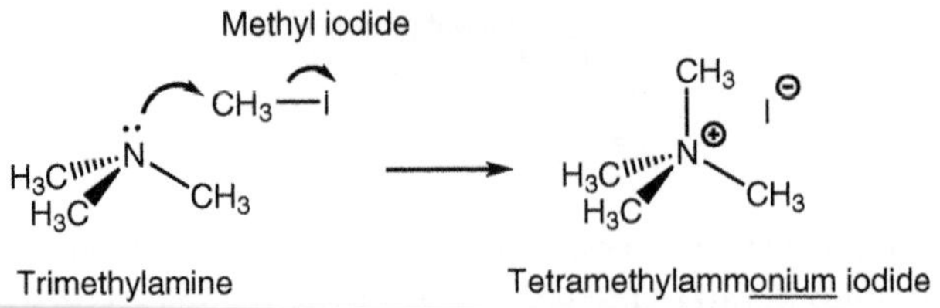

Methyl iodide

Trimethylamine   Tetramethylammonium iodide

Figure 4.18 When a Lewis base (trimethylamine) reacts with a Lewis acid (methyl iodide) and the central atom of the base acquires a positive charge, such species are called "onium salts". Here the resulting compound is named tetramethylammonium iodide.

Unlike Brønsted–Lowry acids and bases, there is no general method of defining the strength of a Lewis acid or base. In general a more electron-deficient central atom will be a stronger Lewis acid. The approximate strength of some Lewis acids is as follows, where X represents a halogen: $BX_3 > AlX_3 > FeX_3 > GaX_3 > SbX_5 > SnX_4 > ZnX_2 > HgX_2$.

CHAPTER 5

Partition Coefficients

WHAT IS A PARTITION COEFFICIENT?

The ability of a drug to reach a specific target in the body depends in part on its ability to cross cell membranes. These are comprised of bilayers of lipid molecules, which have a long nonpolar chain of carbon atoms at one end and a polar group at the other end. *Lipid bilayers* are arranged with the polar groups (head) projecting outward toward the aqueous environment with the nonpolar regions (tail) pointing toward each other (Figure 5.1). Drugs need to have some polar characteristics that impart a degree of water solubility so as to be transported throughout the body. Those polar properties also allow the drug to be attracted to and pass through the polar head groups of lipid bilayers. Once a drug manages to traverse the head groups, however, it is confronted with a highly nonpolar domain that is composed of the hydrocarbon chains of the lipid molecules. A drug that is too polar will have little or no affinity for this region and will not be able to cross. Envision a container of oil and then imagine adding to it a drop of water. The water will bead up and not become dispersed in the oil. Thus compounds such as those containing quaternary ammonium groups have difficulty crossing lipid bilayers because of their high polarity, even though they will often have good water solubility. Most drugs require a balance of polar and nonpolar characteristics so as to ensure that they can readily pass from one bodily compartment to another. The ratio of polar to nonpolar characteristics of a drug is therefore a critical physicochemical property and is known as the *partition coefficient* (P) of the drug.

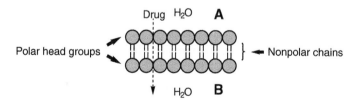

Figure 5.1 This drawing depicts a lipid bilayer. The polar head groups face toward the aqueous environments of compartments **A** and **B**, while the hydrocarbon chains associate to form a nonpolar interior region. For a drug to pass from **A** to **B**, it must possess some polar characteristics to ensure passage through the polar domains and some nonpolar properties to navigate through the interior portion of the bilayer.

Organic Chemistry Concepts and Applications for Medicinal Chemistry
http://dx.doi.org/10.1016/B978-0-12-800739-6.00005-X

The partition coefficient (*P*) is defined as follows:

$$P = [\text{Drug}]_{\text{lipid phase}} / [\text{Drug}]_{\text{water}}$$

Theoretically, one can disperse a drug between equal amounts of a lipid layer and water and then determine the concentration of drug in each layer. The ratio is the partition coefficient. In practice, however, this is not done because the term lipid refers not to a single well-defined compound but to an entire class of molecules. Mixing lipids and water also tends to form emulsions which complicates such determinations. Instead, a well-behaved organic compound is often chosen as a model for the lipid layer. Most commonly this is 1-octanol, a compound with a polar OH group at one end, and a long nonpolar *n*-octyl chain at the other. Thus experimentally it is often *octanol–water partition coefficients* that are determined (Figure 5.2).

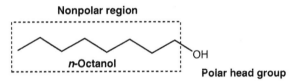

Figure 5.2 The compound *n*-octanol often serves as a model for a lipid molecule in determining partition coefficients.

Octanol–water partition coefficients may be obtained by distributing a drug between *n*-octanol and water in a separatory funnel and, after equilibration, determining the concentration of drug in each layer.

$$P = [\text{Drug}]_{\text{octanol}} / [\text{Drug}]_{\text{water}}$$

The values that are obtained are usually exponential numbers and it is therefore common to express partition coefficients as the logarithm of the partition coefficient, $\log P$. An alternative method employs reverse-phase high-performance liquid chromatography. The chromatographic retention time can be correlated to the $\log P$.

Consider that drug molecules consist of core structures to which a variety of substituents are attached. Each substituent has polar or nonpolar properties that will have an effect on the partition coefficient of the structure onto which it is attached. The quantitative measure of the effect of a substituent on the partition coefficient of the molecule as a whole is called the *Hansch partition coefficients* (π_x) of the substituent.

$$\pi_x = \log\left(P_x / P_H\right) = \log P_x - \log P_H$$

where P_H and P_x are the partition coefficients of the unsubstituted and substituted molecules, respectively.

The values of π_x can be either positive, which means that group x will increase the lipid solubility (*hydrophobicity*) of the compound, or negative, in which case group x will make the compound more water soluble (*hydrophilic*).

The above equation can be rewritten as: $\log P_x = \log P_H + \pi_x$.

Thus the $\log P$ of a compound is equal to the $\log P$ of the parent compound plus the Hansch partition coefficient of the substituent. Alternatively, the partition coefficient can be viewed as being equivalent to the sum of the contributions from all of the various groups from which it is comprised. Therefore,

$$\log P_{compound} = \Sigma\, \pi_x$$

EFFECT OF STRUCTURE ON PARTITION COEFFICIENTS

Consider the compounds benzene and anisole (methoxybenzene). Anisole can be derived from benzene by replacing one of the hydrogen atoms with a methoxy group. It has been reported that the $\log P$ of anisole is 2.11 and that of benzene is 2.13 [1]. Using the equation above, we can calculate a π value for a methoxy group (Figure 5.3).

$$\pi_{(OCH3)} = \log P_{anisole} - \log P_{benzene}$$

$$\pi_{(OCH3)} = 2.11 - 2.13$$

$$\pi_{(OCH3)} = -0.02$$

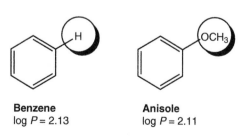

Benzene
log P = 2.13

Anisole
log P = 2.11

Figure 5.3 The Hansch partition coefficient (π_x) of the methoxy group can be derived using partition coefficients of a compound having a methoxy group (such as anisole) and one identical in all respects except for the absence of such a group (benzene).

The replacement of hydrogen with a methoxy group will result in the compound becoming very slightly more water soluble. Once the Hansch partition coefficient of a methoxy group is known, it can be used to calculate the partition coefficients of other compounds. For example, the $\log P$ value of 1,3,5-trimethoxybenzene can be calculated as shown below:

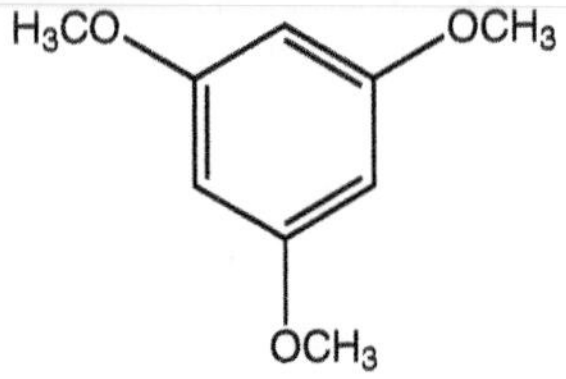

1,3,5-Trimethoxybenzene

$$\log P_x = \log P_{benzene} + 3\pi_{(OCH3)}$$

$$\log P_x = 2.13 + 3\,(-0.02)$$

$$\log P_x = 2.07$$

Examine the structures of acetic acid and methoxyacetic acid (Figure 5.4). Here again we see that the difference between the compounds is replacement of a hydrogen by a methoxy group. But in this case, when the Hansch partition coefficient of the methoxy group is determined, it is seen that the value obtained is considerably more negative ($\pi_{(OCH3)} = -0.24$) than that derived using anisole. One of the differences between methoxyacetic acid and anisole is that in the former the methoxy group is attached to an

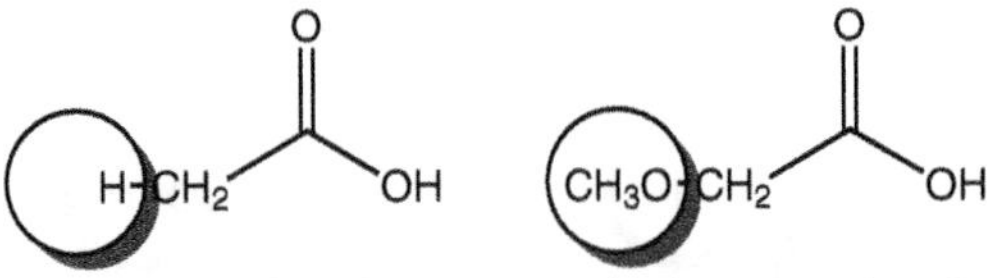

Acetic acid
$\log P = -0.31$

Methoxyacetic acid
$\log P = -0.55$

$$\pi_{(OCH3)} = -0.55 - (-0.31)$$
$$\pi_{(OCH3)} = -0.24$$

Figure 5.4 Using known partition coefficients for methoxyacetic acid and acetic acid, the Hansch partition coefficient of an aliphatic methoxy group can be derived. *The values are obtained from Ref. [1].*

aliphatic carbon while in anisole it is an aromatic carbon to which it is joined. Water solubility depends in part on the ability to form *hydrogen bonds* between a compound and water (Figure 5.5). A lone pair on the ether oxygen supplies the electron density needed to bond with a hydrogen atom of water. An aromatic ring serves as an electron-withdrawing group exerting a pull on the lone pair electrons, effectively decreasing their ability to form hydrogen bonds. There is no such withdrawing effect for an ether group attached to an aliphatic carbon. Therefore, such a compound will more readily form hydrogen bonds with water. This will be reflected in a more negative Hansch partition coefficient for an aliphatic ether as opposed to an aromatic ether. Similar effects are observed with many other functional groups.

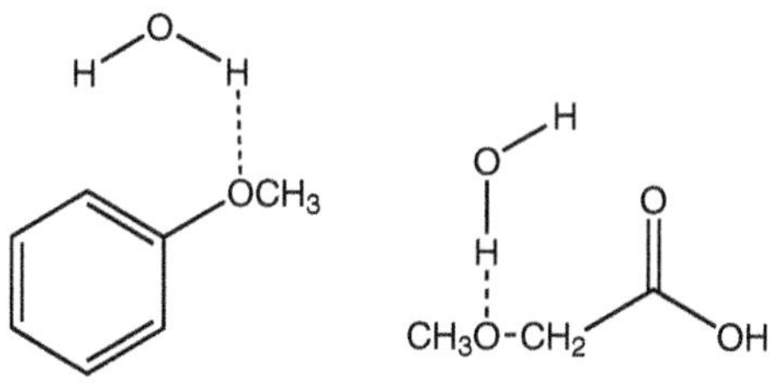

Figure 5.5 The negative values for the Hansch partition coefficients of an aromatic and aliphatic methoxy group are a reflection of the group's ability to hydrogen bond with water. An aromatic ring will withdraw electron density from oxygen reducing its π_x value relative to an aliphatic methoxy group.

Consider now the series of compounds shown in Figure 5.6: acetic acid, propionic acid, and butyric acid. These represent what is known as a *homologous series*, each differs from the preceding compound by one methylene (CH_2) group. Log P values of −0.31, 0.33, and 0.79 respectively have been reported for these compounds [1]. We see that on average, each additional aliphatic carbon adds +0.5 to the log P value. It has been established that when all other factors are equivalent, log P values increase with the number of carbon atoms.

CH_3—CO_2H	log P = −0.31	
		Δlog P = 0.64
CH_3CH_2—CO_2H	log P = 0.33	
		Δlog P = 0.46
$CH_3CH_2CH_2$—CO_2H	log P = 0.79	

Figure 5.6 Partition coefficients for the homologous series acetic acid, propionic acid, and butyric acid. On average, each additional aliphatic carbon adds 0.5 to the partition coefficient. *Values are from Ref. [1].*

The compounds n-butanol and i-butanol are positional isomers ($C_4H_{10}O$) that differ only by the presence of a *branch* in i-butanol (Figure 5.7). If one subtracts the $\log P$ value of n-butanol from that of i-butanol, the difference is the Hansch partition coefficient for the branch. On average, this value is -0.2, meaning that branching tends to increase water solubility to a small extent.

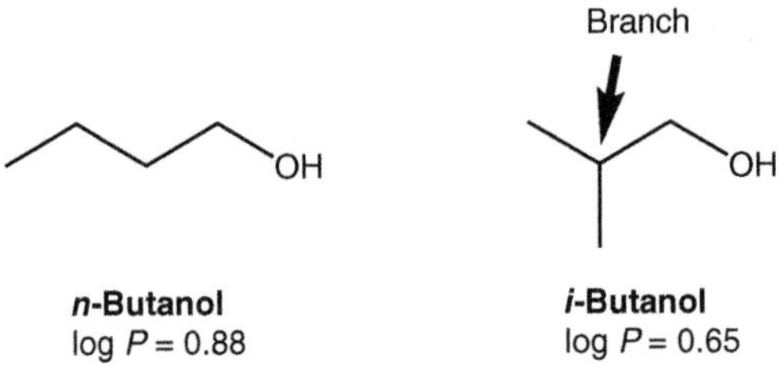

Branch

n-Butanol
$\log P = 0.88$

i-Butanol
$\log P = 0.65$

$\pi_{branch} = \log P_{i\text{-BuOH}} - \log P_{n\text{-BuOH}}$

$\pi_{branch} = 0.65 - 0.88$

$\pi_{branch} = -0.23$

Figure 5.7 The structures of n-butanol and i-butanol showing that the latter has a branch in the alkyl chain. A Hansch partition coefficient is derived here for a branch. *The values are from Ref. [1].*

To understand why branching lowers the partition coefficient, consider what happens when a more extended compound such as n-butanol and a branched compound (i-butanol) partition from an organic solvent such as octanol into water. Figure 5.8 shows that the critical volume (the volume occupied by 1 mol of a compound) calculated for i-butanol is smaller than that of n-butanol.

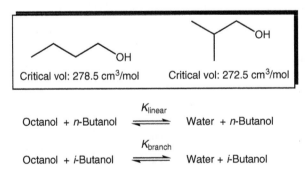

Critical vol: 278.5 cm³/mol   Critical vol: 272.5 cm³/mol

$$\text{Octanol} + n\text{-Butanol} \underset{}{\overset{K_{linear}}{\rightleftharpoons}} \text{Water} + n\text{-Butanol}$$

$$\text{Octanol} + i\text{-Butanol} \underset{}{\overset{K_{branch}}{\rightleftharpoons}} \text{Water} + i\text{-Butanol}$$

Figure 5.8 Critical volumes for n-butanol and i-butanol as calculated using ChemBio-Draw Ultra 13.0, Perkin Elmer Informatics, Waltham, MA. Partitioning behavior can be viewed as an equilibrium between the substrate in dissolved in octanol and water.

An equilibrium can be established between n-butanol in octanol and n-butanol in water with a constant called K_{linear}. A similar equilibrium can be written for i-butanol with constant K_{branch}. As the organic substrates begin to partition into water, the water molecules must organize around the organic molecules. The more extended compound n-butanol will require more water molecules to organize than i-butanol because it occupies a larger volume. A higher degree of organization means lower *entropy* (S) and so $\Delta S_{linear} < \Delta S_{branch}$. The change in entropy for a process is related to the equilibrium constant by two equations. The first of these states that as ΔS decreases, the free energy change (ΔG) for a process increases.

$\Delta G = \Delta H - T\Delta S$, where ΔH is the change in enthalpy, T is the temperature in Kelvin (K) and ΔS is the entropy change

A second equation relates ΔG to the equilibrium constant.

$\Delta G = -RT \log K$, where R is the universal gas constant and T is the temperature (K)

Thus $-\Delta G$ is proportional to $\log K$ and as ΔG increases K decreases. The difference in ΔS for a linear vs a branched compound therefore translates into K_{branch} being larger than K_{linear}. A branched compound will partition into water to a greater extent than a linear compound and will therefore have a smaller partition coefficient.

Another structural feature that has an effect on partition coefficient is unsaturation. Figure 5.9 shows the calculated $\log P$ values for cyclohexane, cyclohexene, and benzene. The difference between cyclohexene and cyclohexane is the replacement of two hydrogens by a π-bond leading to a significant decrease in the partition coefficient. Thus π-bonds can also be considered to have negative values for their Hansch partition coefficients. Benzene, with three π-bonds, has an even lower partition coefficient than cyclohexene. Remember that electrons in π-bonds are not held as tightly by the nuclei as those in σ-bonds. They can be donated to a certain extent to form hydrogen bonds with water, which thereby increases water solubility.

log P = 3.35   log P = 2.87   log P = 2.13

Figure 5.9 Calculated log P values for cyclohexane, cyclohexene, and benzene showing that unsaturation lowers the partition coefficient. Values were calculated using Chem-BioDraw Ultra 13.0, Perkin Elmer Informatics, Waltham, MA.

A listing of some structural effects that decrease the partition coefficient (have negative values of π_x) includes the following:

- Any functional group that can readily form hydrogen bonds with water. This includes carbonyl-based groups and other oxygen and nitrogen-containing functionality such as alcohols, ethers, amines, nitro groups, nitriles, sulfonamides, etc.
- Charged groups such as carboxylate anion, alkoxide anion, and ammonium groups. Salts also fall into this category. Positive charges are attracted to the oxygen of water while negative charges are attracted to the hydrogens.
- Unsaturation
- Branching
- Isolated aliphatic fluoro groups. Fluorine has three lone pairs of electrons and is a small element, which facilitates the formation of H bonds with water.

Groups that do not readily form hydrogen bonds with water will generally increase the partition coefficient (have positive values for π_x). Among these are the following:

- Carbons (CH_3, CH_2, CH, C). Unsaturated carbons (sp^2 and sp) will have less of an effect than aliphatic carbons but will still increase the $\log P$.
- Halogens: $-I, -Br, -Cl$, and $-F$ if it is attached to an unsaturated carbon. The electron-withdrawing effect of a multiple bond removes electron density from fluorine making the formation of hydrogen bonds more difficult. The other halogens are all much larger than water rendering hydrogen bond formation more difficult.
- Trifluoromethyl (CF_3): Three strongly electronegative fluorines attached to one carbon exert a strong polar effect on each other's lone pair electrons decreasing their ability to form hydrogen bonds. The CF_3 group is one of the strongest $+\pi$ groups.
- Thiols and sulfides: The larger size of sulfur as compared to water decreases its ability to hydrogen bond with water.

REFERENCE

[1] A. Leo, C. Hansch, D. Elkins, Partition coefficients and their uses, Chemical Reviews 71 (1971) 525–616.

CHAPTER 6

The Nomenclature of Cyclic and Polycyclic Compounds

When organic chemistry was in its infancy, compounds were named by the individual who first isolated or synthesized them. Such names often reflected either properties of the compound or the source of the compound, or were simply the result of the chemist's imagination. When the total number of known compounds is small, such a system of nomenclature is manageable. A great many of these so-called trivial names have become so firmly entrenched that attempts to change them to more systematic names have met with stiff opposition. How many chemists, for example, refer to acetic acid as ethanoic acid? Despite the preference for certain trivial names, there is a need for a standard system of naming organic compounds. Both Chemical Abstracts Service (CAS) and the International Union of Pure and Applied Chemistry (IUPAC) have devised systems of nomenclature that have gained widespread acceptance. These systems are similar in many respects and no attempt will be made in this chapter to distinguish them. What is important to realize is that systematic nomenclature unambiguously defines the carbon skeleton, the placement of substituents and heteroatoms, and any stereochemical information about a given compound. Drugs are typically referred to by three different names—a systematic name, a generic name, and a trade name. For example, N-(4-hydroxyphenyl)acetamide is the systematic name associated with the generic drug names of acetaminophen or paracetamol. This compound is sold under the trade names of Tylenol®, Apra®, and Feverall® to name just a few. Of these, only the systematic name will never change. In this chapter, the focus will be on learning the rules of nomenclature so as to be able to draw a correct chemical structure for a complex molecule when provided with a systematic name.

COMMON HETEROCYCLES

A perusal of any reference that lists the chemical structures of those drugs that are most often prescribed will reveal that the majority include one or more heterocyclic rings. This is not a coincidence. When a heteroatom is introduced into a ring, the electronegativity difference between the heteroatom and carbon induces partial charge development which in turn will affect how that ring interacts with a receptor. Lone pairs on the heteroatom(s) also serve as

Organic Chemistry Concepts and Applications for Medicinal Chemistry
http://dx.doi.org/10.1016/B978-0-12-800739-6.00006-1

sites for hydrogen bonding interactions. Thus, knowledge of those heterocycles that are commonly found in drugs is as much a prerequisite to drawing drug structures from systematic names as learning the alphabet is to writing an essay.

Figure 6.1 shows the structures for 32 five-membered heterocyclic rings and immediately raises concerns about how one can possibly learn to associate their names and structures. The task is not as daunting as it seems. Of these compounds, 30 contain nitrogen and special rules apply for their names. Examine Figure 6.1 again and focus on the compounds in the left column of the first six rows. The first part of their names indicates the pattern of hetero-atoms in the ring and the -ole ending signifies the presence of two double bonds. In each case, there is one heteroatom that is not double bonded. It uses a lone pair to make those rings aromatic (Chapter 1). The top three compounds on the left-pyrrole, pyrazole, and imidazole have trivial names. Pyrrole has just one nitrogen while pyrazole and imidazole each have two. In pyrazole, the nitrogens are adjacent while in imidazole, they are separated by one carbon. When there are two different heteroatoms, they are named systematically. Thus oxazole has an oxygen (ox) and a nitrogen (az) and thiazole has a sulfur (thi) and a nitrogen (az). If the name is preceded by is- (if the next letter is a vowel as in isoxazole) or iso- (if the next letter is a consonant as in isothiazole, not shown), the heteroatoms are adjacent. If not, they are separated by a carbon. These six rings then provide the basis for learning all 30 of the nitrogen-containing rings. Focus now on those compounds in the right column. Notice that in each case, the pattern of heteroatoms is the same as for the compound on the left, as is the first part of the name, but the ending is different. Those on the right have no double bonds in the ring (they are saturated rings) and their names all end with -olidine. Thus, **pyrrole** has one nitrogen in the ring and two double bonds, whereas **pyrrolidine** has one nitrogen and no double bonds. Now examine the compounds in the three inside columns. In each instance, they have only one double bond and they all end with -oline. Again it is the first portion of the name that indicates the pattern of heteroatoms. There are, however, several possible isomers of these monounsaturated heterocycles. They are differentiated by indicating the starting position of the double bond by number, with the understanding that it extends to the next highest numbered position. Thus 2-pyrroline has one nitrogen in the ring and one double bond between the 2- and 3-positions. This brings up the question of how to number the positions in these rings. The rules are as follows:

- A single heteroatom in a ring becomes the 1-position and the remainder of the ring is numbered consecutively in a clockwise (or counterclockwise) direction.

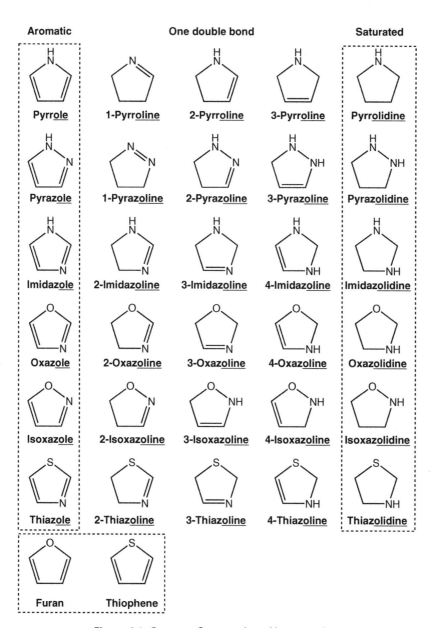

Figure 6.1 Common five-membered heterocycles.

- For two identical heteroatoms, one is the 1-position and the other is the next possible higher number. Thus in pyrazole, one nitrogen is at position 1 and the other is 2 with numbering then continuing around the ring. In imidazole, one nitrogen is position 1 and the other is the 3-position.
- If one of the heteroatoms is saturated, then it is assigned the lower number. Thus it is the saturated nitrogen of imidazole (the one without any double bonds) that is position 1.

- If the heteroatoms are different as in oxazole, then there is a heteroatom preference rule that states that the atom in the higher group number of the Periodic Table has preference (takes the lower number). For two atoms of the same group, the one with the lower atomic number takes preference. Stated simply, for those heteroatoms most likely found in drug molecules, the order of preference is $O > S > N > P$.

The remaining two rings at the bottom of Figure 6.1 do not follow the same conventions of name endings as the nitrogen-containing five-membered rings. They are given the trivial names of furan and thiophene and are both aromatic. Analogs with fewer double bonds are named systematically according to rules that will be given later in this chapter.

Five-membered rings with three or more heteroatoms are all named systematically (Figure 6.2). The designation di-, tri-, or tetra- is used to indicate the presence of two, three, or four of a particular heteroatom. Numbers precede the name when necessary to indicate, in order, the location of the heteroatoms. As long as the ring contains nitrogen, the rules about name endings apply to indicate the number of double bonds.

A series of six-membered heterocyclic rings is shown in Figure 6.3. Those in the top row are aromatic and are given trivial names. We have already encountered pyridine in Chapter 1. Pyridazine, pyrimidine, and

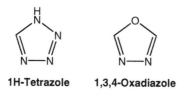

1H-Tetrazole   1,3,4-Oxadiazole

Figure 6.2 Systematically named five-membered heterocycles.

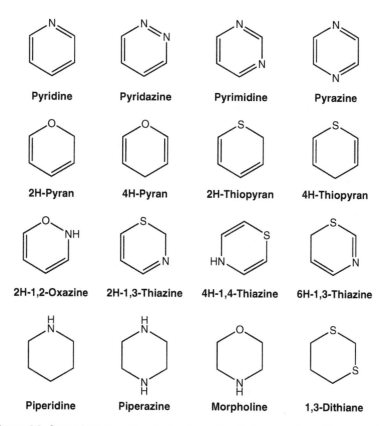

Pyridine	Pyridazine	Pyrimidine	Pyrazine
2H-Pyran	4H-Pyran	2H-Thiopyran	4H-Thiopyran
2H-1,2-Oxazine	2H-1,3-Thiazine	4H-1,4-Thiazine	6H-1,3-Thiazine
Piperidine	Piperazine	Morpholine	1,3-Dithiane

Figure 6.3 Some common saturated and unsaturated six-membered heterocycles.

pyrazine all have two nitrogens in the ring with the heteroatoms located at the 1,2-, 1,3-, and 1,4-positions, respectively. The DNA bases cytosine and thymine are derived from pyrimidine. The saturated nitrogen-containing rings in the bottom row are given the trivial names of piperidine, piperazine, and morpholine. They are prevalent in drug molecules. A six-membered ring that has one oxygen is called pyran and its sulfur analog is thiopyran. These rings contain the maximum degree of unsaturation that is possible. A six-membered ring can generally accommodate three double bonds. Neither oxygen nor sulfur however can be part of a double bond in a ring. Each is divalent and both valencies are used to form the ring. If a double bond formed from oxygen or sulfur, they would be trivalent, and would be positively charged. That is an unstable arrangement. In an odd-sized ring such as a five- or seven-membered ring, oxygen or sulfur can be the lone saturated atom when the ring has its maximum allotment of double bonds. In an even-sized ring such as a six- or eight-membered ring, the presence of oxygen or sulfur within the ring means that there will be one fewer double bond than the maximum number. Thus one other atom in the ring besides the heteroatom must be saturated. That position is indicated by a number followed by a capital H preceding the name as in 2H-pyran. Thus, the oxygen and 2-position are saturated and there are two double bonds, one at the 3,4- and the other at the 5,6-positions. Most six-membered rings containing two or more heteroatoms are named systematically. Note that the name generally ends with -ine and that the rings are assumed to have the maximum amount of unsaturation.

Rings of 7–10 members are named systematically, but using a clever device to indicate ring size. Consider that heptane is the name of a seven-carbon aliphatic compound. The first vowel–consonant combination that occurs in heptane is "ep". Thus, a seven-membered heterocycle has a name ending with -epine. Table 6.1 lists the designations for 7–10 membered heterocyclic rings.

Table 6.1 Nomenclature of Medium-Sized Rings

Ring Size	Hydrocarbon	Ring-Size Designator
7	Heptane	-epine
8	Octane	-ocine
9	Nonane	-onine
10	Decane	-ecine

The smallest rings consisting of three or four atoms have no special rules for their naming. Figure 6.4 shows the structure and name for several of the more common small rings.

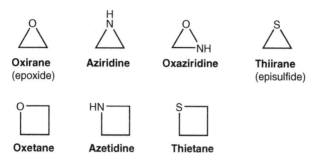

Figure 6.4 Several small-ring heterocycles.

Rules for systematically naming and numbering fused polycyclic rings will be thoroughly reviewed in the next section of this chapter, but there are several such ring systems that are common and have been given trivial names. Some bicyclic rings are shown in Figure 6.5. Note that in the compounds quinoline and 1H-indole, the nitrogen is located at the 1-position. Isomers in which the nitrogen is located at the 2-position are called isoquinoline and 2H-isoindole.

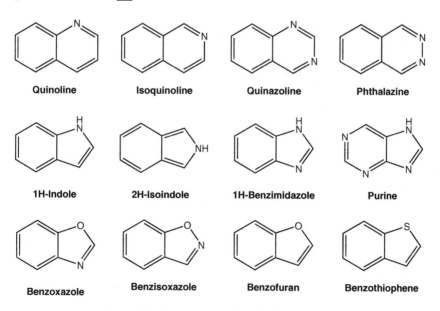

Figure 6.5 Common fused bicyclic heterocycles.

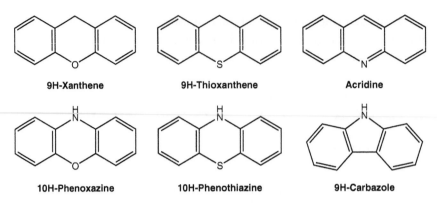

Figure 6.6 Some fused tricyclic heterocycles.

A few tricyclic rings containing heteroatoms are listed in Figure 6.6. Of these, the best known is 10H-phenothiazine, which forms the basis of an entire class of drugs. For completeness, the structures of common hydrocarbon rings are given in Figure 6.7.

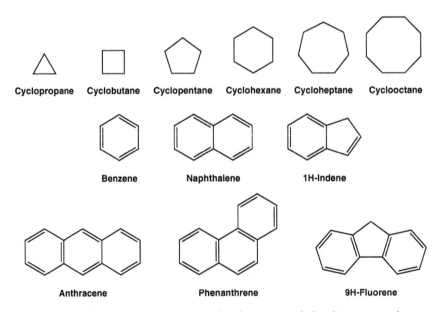

Figure 6.7 Several common saturated and unsaturated alicyclic compounds.

FUSED-RING SYSTEMS

Two rings that share one or more bonds are said to be fused. The result of fusing two smaller rings is the creation of a new, larger ring system. Many drug structures incorporate fused-ring systems and it is therefore important to understand the systematic nomenclature and numbering of such rings.

In the previous section, we reviewed the structure and numbering of those rings that will serve as the building blocks for more complex fused-ring systems such as pyrido[2,3-c]pyridazine. This is a typical example of what is known as *prefix–suffix nomenclature.* In this nomenclature convention, the term "pyrido" is the prefix and pyridazine is the suffix ring. The prefix is usually derived from the name of a ring by replacing its ending with "o". Thus pyrid<u>o</u> comes from pyrid<u>ine</u>. Of the two rings, the prefix ring is the smaller or less complex. In this example, a new ring system is being created by fusing a pyridine and a pyridazine ring. Since not every bond in either ring is identical, we need to know which bonds to fuse together. This is the job of the bracket term [2,3-c]. The numbers 2 and 3 refer to positions on the prefix ring; in this case, the 2- and 3-positions of pyridine. The letter "c" refers to a particular bond in the suffix ring (Figure 6.8). The bond between the 1- and 2-positions of any ring is labeled as the "a" bond. Lettering then continues in sequence around the entire ring, with the bond between the 2- and 3-positions being "b", and so on.

Once the rings and positions involved in the fusion have been identified, the rings should be brought into proper alignment to allow the fusion to proceed. It is advisable at this point to draw the rings without their double

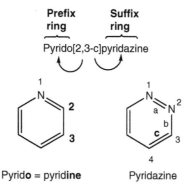

Figure 6.8 An example of prefix–suffix nomenclature of fused aromatic compounds. Note that the numbers within the bracket refer to positions on the prefix ring, whereas the letters are associated with bonds within the suffix ring.

bonds to avoid problems with incorrect valencies later. Remember however, that the rings being fused are unsaturated and so the fused–ring system will have the maximum amount of unsaturation that will be added back later. The order of the numbers within the bracket term is important. The position on the prefix ring corresponding to the first number will end up joined to the lower numbered position of the lettered bond. Then the last number will join to the higher numbered position of the lettered bond. The rings are then joined. This sounds a little confusing but is demonstrated in Figure 6.9. After the fusion, the original rings should still be evident in the new ring system (dashed circles). If not, then a mistake was made during the fusion.

The new ring system has now been created. If the name does not specify that any positions are saturated, then the maximum amount of unsaturation will be added back into the new ring system (Figure 6.10). For two fused six-membered rings, the maximum number of double bonds is five. Make sure when drawing double bonds that you do not exceed the normal valency for any atoms in the ring. Thus there can be no more than four bonds to carbon, three to nitrogen (if uncharged), or two to oxygen or sulfur.

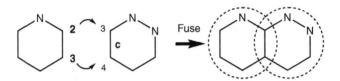

Figure 6.9 The order of the numbered positions indicates how the fusion will occur with the lettered bond. Once the fusion has been completed, the original rings should still be observed within the new ring system. If not, the fusion was performed incorrectly.

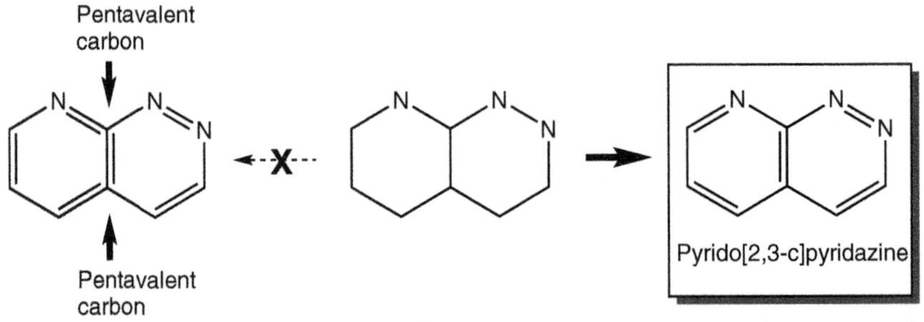

Pentavalent
carbon

Pentavalent
carbon

Pyrido[2,3-c]pyridazine

Figure 6.10 Once the fused-ring system has been created, the maximum number of double bonds is added back into the ring system. Remember carbon can have no more than four, nitrogen three, and oxygen and sulfur two bonds.

Now that the new ring system has been created, every position must be numbered so that substituents can be attached at the proper locations. As it turns out, this process is a little more involved than it was for the simple rings that were discussed in the first section of this chapter.

For a polycyclic fused-ring system, numbering could vary depending on how one orients the structure on the page. It is therefore important that the ring be restricted to one proper orientation before its positions can be numbered. There are a series of rules that must be applied in sequence for orienting such molecules.

Orientation Rule 1. The ring system must be oriented such that the greatest number of adjacent rings forms a horizontal line.

Orientation Rule 2. After applying Rule 1, the molecule is oriented so that the greatest number of remaining rings is above and to the right of the horizontally oriented rings (Figure 6.11).

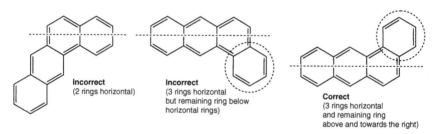

Incorrect
(2 rings horizontal)

Incorrect
(3 rings horizontal but remaining ring below horizontal rings)

Correct
(3 rings horizontal and remaining ring above and towards the right)

Figure 6.11 Fused-ring systems must be properly oriented before the positions can be numbered. The first two rules, which must be applied in order, indicate that the maximum number of fused rings must form a horizontal array. After that additional rings should be located above the horizontal rings if possible and toward the right.

Numbering begins in the right-hand or upper right-hand ring. The most counterclockwise position that is not a ring junction (an atom common to two or more rings) becomes the 1-position. Numbering then continues in a clockwise direction around the entire structure at all non-ring-junction positions. If a heteroatom occurs at a ring-junction position, it takes the next number in sequence. At all other ring-junction positions the number of the position that immediately precedes it is assigned, followed by the lowest possible letter. If a ring junction occurs immediately after another ring junction, it takes the same number and the next letter. Ring junctions that occur in the interior of a ring system are assigned the highest number–letter sequence moving in a clockwise direction (Figure 6.12).

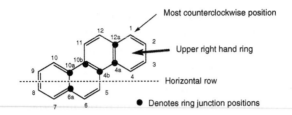

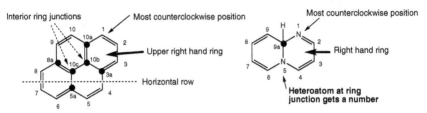

Figure 6.12 Once the ring has been oriented, it can be numbered. Numbering begins in the right-hand or upper right-hand ring and then continues clockwise around the entire system. Ring junctions take number–letter combinations. A heteroatom that occurs at a ring junction takes the next highest number.

Orientation Rule 3. In those instances in which a heteroatom is part of the ring system, first orient the molecule as in Rules 1 and 2 and then so that the heteroatom(s) occurs at the lowest possible number(s) (Figure 6.13).

Orientation Rule 3a. If after applying Rule 3 there is a choice between two or more heteroatoms, orient so that the heteroatom of higher preference (O > S > N > P) is at the lowest possible number (Figure 6.14).

Orientation Rule 4. If after applying Rules 1–3a there is still ambiguity, orient the molecule so that the ring junctions occur at the lowest possible numbers (ignore the letters) (Figure 6.15).

Orientation Rule 5. If ambiguity remains after applying Rules 1–4, orient the molecule so that saturated atoms have the lowest possible numbers (Figure 6.16).

Orientation Rule 6. In the event that there is still a choice after applying Rules 1–5, orient the molecule so as to have substituents located at the lowest possible position (Figure 6.17).

Although the vast majority of polycyclic ring systems are oriented and numbered using the rules listed above, there still remain a few rings that are numbered nonsystematically. These numbering schemes, much like trivial names, have become accepted over the years and are shown in Figure 6.18. Anthracene and the linear tricyclic analogs acridine, 9H-xanthene, and 9H-thioxanthene are numbered beginning properly in the right-hand ring but then skip to the ring on the left and finish with the 9- and 10-positions

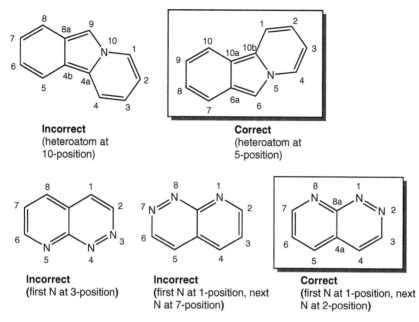

Incorrect
(heteroatom at
10-position)

Correct
(heteroatom at
5-position)

Incorrect
(first N at 3-position)

Incorrect
(first N at 1-position, next
N at 7-position)

Correct
(first N at 1-position, next
N at 2-position)

Figure 6.13 After Rules 1 and 2 have been applied, if alternative orientations are still possible, Rule 3 states that the molecule is oriented so that heteroatoms occur at the lowest possible number(s).

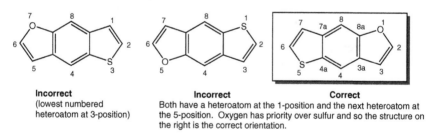

Incorrect
(lowest numbered
heteroatom at 3-position)

Incorrect
Both have a heteroatom at the 1-position and the next heteroatom at the 5-position. Oxygen has priority over sulfur and so the structure on the right is the correct orientation.

Correct

Figure 6.14 Rule 3a, which states that the order of preference of heteroatoms is O > S > N > P, is invoked if Rules 1–3 still allow for alternative orientations.

in the middle ring. Phenanthrene is numbered starting in the right-hand ring but in a direction opposite to that which is deemed proper. Again, numbering jumps to the ring on the left and is completed in the middle ring. 9H–Carbazole is a nitrogen analog of the hydrocarbon fluorene and is numbered the same as fluorene, putting the nitrogen at the highest numbered position. Finally, the bicyclic compound purine is numbered completely around the pyrimidine ring in a counterclockwise manner and then around

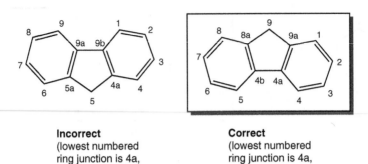

Incorrect
(lowest numbered
ring junction is 4a,
next is 5a)

Correct
(lowest numbered
ring junction is 4a,
next is 4b)

Figure 6.15 If ambiguity remains after applying Rules 1–3a, Rule 4 states that the molecule be oriented so that ring junctions occur at the lowest possible numbers.

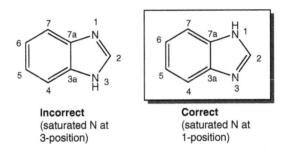

Incorrect
(saturated N at
3-position)

Correct
(saturated N at
1-position)

Figure 6.16 After applying Rules 1–4, if the orientation is still in question, arrange the molecule so that any saturated position(s) occur(s) at the lowest possible number(s).

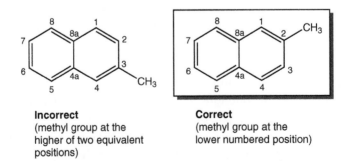

Incorrect
(methyl group at the
higher of two equivalent
positions)

Correct
(methyl group at the
lower numbered position)

Figure 6.17 The final rule, invoked if there is still a choice after applying Rules 1–5, is to orient the molecule so that substituents occur at the lowest possible positions.

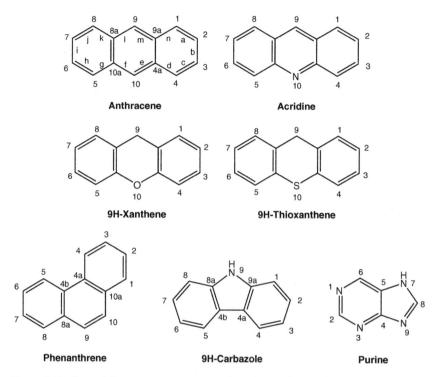

Figure 6.18 Several ring systems are not numbered according to the rules stated above. The structures and numbering of these rings is shown here.

the imidazole ring in a clockwise direction. It is important to realize however when labeling bonds in these rings that the "a" bond connects the 1- and 2-positions and bond lettering then continues in the direction of the higher numbered position for each bond in sequence. As the reader will see later in this chapter, many natural products such as the steroids and morphinans have their own unique numbering systems.

Now that we have gone through the process of drawing a fused-ring system and know how to properly orient the molecule and number each position, let us examine a few additional examples. With pyrrolo[3,2,1-kl] phenothiazine (Figure 6.19), we have a situation not encountered to this point, namely three numbers and two letters within the bracket term. All this means however is that three positions on the prefix ring are to be fused to two bonds on the suffix ring. The order of the numbers tells us that the 3-position of a pyrrole is to be fused to the lower numbered position of the k-bond of phenothiazine while the 1-position of the pyrrole is fused to the higher numbered position of the l-bond of phenothiazine. One need not

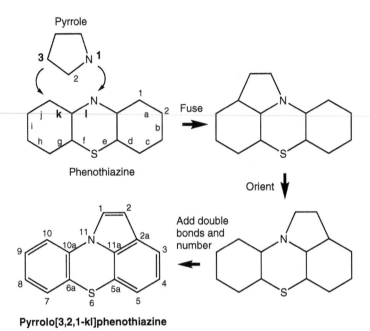

Pyrrolo[3,2,1-kl]phenothiazine

Figure 6.19 Pyrrolo[3,2,1-kl]phenothiazine provides an example of a compound in which three positions on the prefix ring are fused to two bonds on the suffix ring.

worry about the 2-position of pyrrole because it will fall nicely into place between the k- and l-bonds. Again, it is helpful to temporarily remove the unsaturation from the rings to be fused. After performing the fusion, the new molecule must now be oriented so that it can be numbered properly. Rules 1 and 2 are used to align three rings horizontally with the five-membered ring above and skewed to the right. Now the unsaturation can be returned to the new ring system (neither nitrogen which has three bonds within the ring system nor sulfur will take part in any double bonds) and the ring can be numbered, starting with the position in the five-membered ring adjacent to the ring-junction nitrogen.

The next example, pyrazino[2,1-a]pyrido[2,3-c][2]benzazepine, is of a fusion of three different rings. This ring system is highly instructive and is shown in Figure 6.20. There are two prefix rings, pyrazine and pyridine, and each has a bracket term indicating its attachment to the one suffix ring, [2] benzazepine. Within the brackets, the numbers always refer to positions on the prefix ring and the letter(s) to bonds of the suffix ring. Let us focus for a moment on the suffix ring in this example. Why is there a single number in brackets preceding the name? As it turns out, the suffix ring is actually

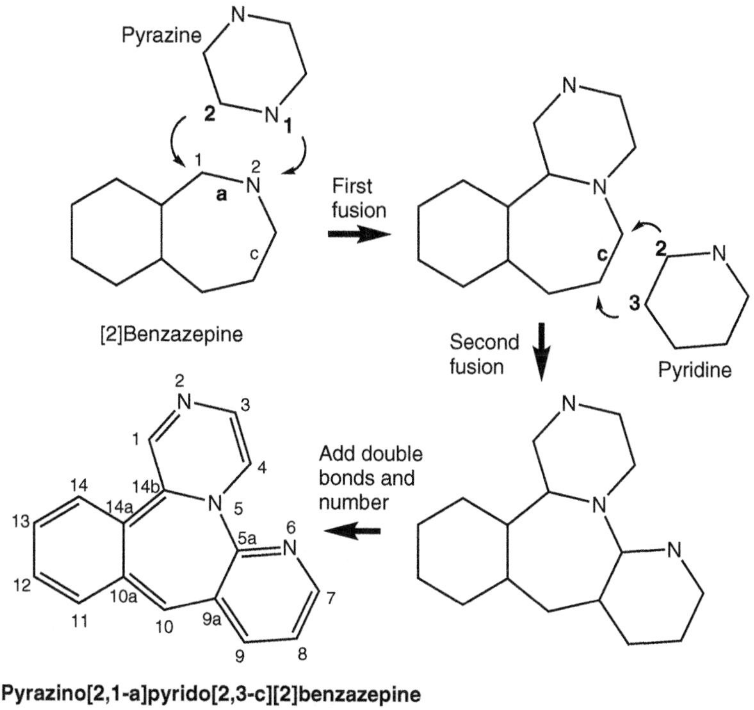

Pyrazino[2,1-a]pyrido[2,3-c][2]benzazepine

Figure 6.20 Pyrazino[2,1-a]pyrido[2,3-c][2]benzazepine is a compound in which two different prefix rings, pyrazine and pyridine, are fused to two different bonds of the suffix ring, in this case [2]benzazepine. Each fusion has its own bracket term.

itself a fused-ring system derived from a benzene and an azepine ring. Benzene is a completely symmetric ring in which every position and bond is equivalent—hence there is no need to specify its points of attachment. An azepine ring is a seven-membered ring containing a single nitrogen atom. If we mentally replace the nitrogen with a carbon, the result is a cycloheptane ring which like benzene has all equivalent bonds. In this nomenclature convention, we simply fuse a six- and a seven-membered ring together to create a new ring system, but then the nitrogen that was mentally removed must be restored into the seven-membered ring. However, the positions are no longer equivalent and so the location of the nitrogen must be specified. That is the purpose of the [2]. Returning now to our example, we see that pyrazine is fused to the a-bond of [2]benzazepine, with the 2-position of pyrazine fused to the 1-position of the suffix ring and the 1-position (N) of pyrazine fused to the 2-position. We see also from the name that pyridine is fused to the c-bond of [2]benzazepine with the pyridine 2-and 3-positions attached to the 3- and 4-positions, respectively, of the suffix ring. As it turns out if one begins this process with the [2]benzazepine ring properly oriented, the resulting fused-ring system will also be correctly oriented with the pyrazine ring above the remaining rings.

A very common situation that arises involves having two benzenes fused to another ring. Since each position of benzene is equivalent, it is not necessary to specify those positions within the bracket term. Instead only the letters of the bonds within the suffix ring are needed. As an example, consider 5H-dibenzo[a,d]cycloheptene (Figure 6.21). In this name, the 5H term indicates that after the fusion, the resulting compound will have no double bonds to the 5-position within the ring system. The prefix rings are two (di-) benzenes and they will be fused to the a- and d-bonds of a seven-membered ring. The ending of the suffix, -ene, indicates that this carbocyclic ring will be unsaturated. Again, remove double bonds initially and designate any bond of the resulting cycloheptane as the a-bond and then locate the d-bond. Now fuse benzene rings to those two bonds. The resulting tricycle will be linear (within reason) and so jump to orientation Rule 4 to arrange the molecule so that the ring junctions fall at the lowest possible numbered positions. The ring can now be numbered. Remember that no double bond will extend to or from the 5-position and so install as many double bonds as possible at all other positions.

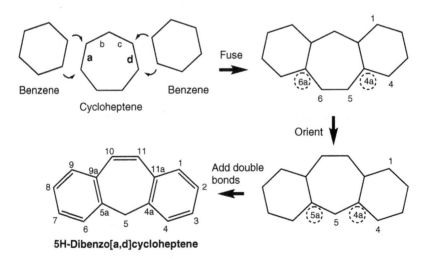

5H-Dibenzo[a,d]cycloheptene

Figure 6.21 The term dibenzo indicates that two benzene rings are fused to a suffix ring. Since every position on benzene is identical, there is no need for numbers within the bracket term.

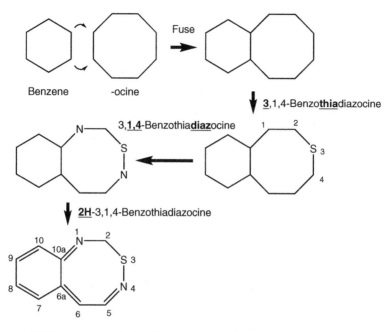

Figure 6.22 When benzene is fused to a nonaromatic heterocycle, the fusion of rings of the proper size is performed and the location of the heteroatoms is indicated in order of occurrence, before the name.

When a benzene is fused to a nonaromatic heterocycle, the numbers that indicate the position of the heteroatom(s) in the fused-ring system often precede the name as in 2H-3,1,4-benzothiadiazocine (Figure 6.22). In this example, we have a benzene fused to an eight-membered (-ocine) ring. Once the fusion has occurred, the name indicates that the first heteroatom (thia) be inserted into the 3-position with two nitrogens (diaz) placed at the 1- and 4-positions, respectively. Here we have sulfur in an even-sized ring which requires that there must be one other saturated position, indicated by the 2H. All other positions will be unsaturated.

SUBSTITUENTS AND SATURATION

The preceding sections have hopefully enabled the reader to draw and number a large variety of cyclic and polycyclic structures. Usually, when dealing with drugs, these structures are modified and "decorated" with a variety of functional groups that play a role in determining their medicinal activity. These modifications and additions to the basic structure are always described either at the beginning and/or at the end of the name of the compound. Several examples are shown below with the parent compound underlined:

- 3,4-Dihydro-1-methyl-1H-2-benzopyran
- 3,4-Dihydrophenanthren-1(2H)-one
- Pyrido[2,3-c]pyridazine, 5-chloro-7,8-dihydro-8-methyl-

The reader may notice that each of these compound names contains the term "dihydro". Remember from earlier in this chapter when a single saturated position in a ring was designated with a number and a capital H, as in 1H-. Hydrogenation of a double bond results in the addition of hydrogen to each end of the π-bond giving rise to a saturated compound. The term "dihydro" is therefore used whenever a pair of positions (although not necessarily adjacent) becomes saturated. Thus dihydro indicates that the compound has one fewer double bond than the parent compound and the numbers that precede "dihydro" give the location of those saturated positions. Note that dihydro does not necessarily mean that hydrogen is added to those positions, only that they no longer have double bonds within the ring. If the equivalent of two, three, or four double bonds is saturated, then tetrahydro-, hexahydro-, and octahydro- are employed again with numbers indicating which positions are saturated. If a compound has three saturated positions, then a combination of H and dihydro, such as 1H-2,3-dihydro-, is used to indicate that fact. For five saturated positions, it would be 1H-2,3,4,5-tetrahydro, etc. The terms trihydro-, pentahydro- heptahydro-, etc. are never acceptable (Figure 6.23).

This knowledge will now allow us to draw the structure of 3,4-dihydro-1-methyl-1H-2-benzopyran (Figure 6.24). Begin by drawing the

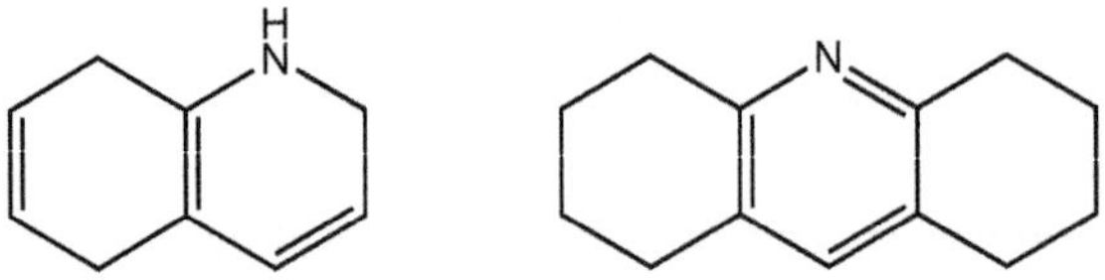

1,2,5,8-Tetrahydroquinoline   1,2,3,4,5,6,7,8-Octahydroacridine

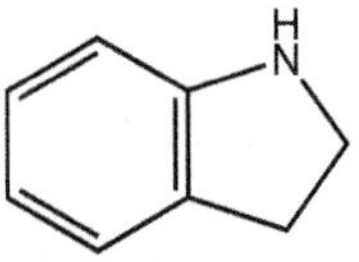

1H-2,3-Dihydroindole
NOT 1,2,3-Trihydroindole

Figure 6.23 When the equivalent of a double bond is saturated, it is indicated by the term dihydro. A single saturated position is designated with its number and then H.

2-benzopyran ring system, omitting for the moment any double bonds. Since oxygen is found in an even-sized ring, we know that there will be one other saturated position, which in this case is the 1-position (1H). The name also indicates that the 3- and 4-positions are saturated (3,4-dihydro). Double bonds can now be added at all other bonds. Attachment of a methyl group to the 1-position completes the structure.

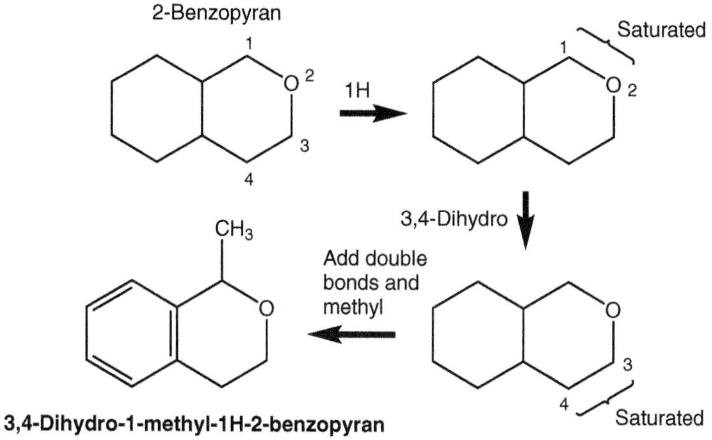

Figure 6.24 The stepwise process used to draw the structure of 3,4-dihydro-1-methyl-1H-2-benzopyran.

A carbonyl group can occur in a structure either with the carbon as part of the basic ring system or attached to the ring system by a single bond, and each situation is named differently. When the carbonyl carbon is a member of the ring, it is designated either within the substituent list as **#-oxo** (e.g. 5-oxo) or by using –**one** as a suffix to the name (Figure 6.25). It is important not to confuse oxo- with oxa- which indicates that an oxygen is a member of a particular ring system. Alternatively, a carbonyl can be attached to a ring system by a single bond. In this instance, the carbonyl group is listed among the substituents and is designated by taking the name of the carboxylic acid from which it derives, dropping the "–ic acid" portion of the name and replacing it by "–yl". Thus an acetyl group is derived from acetic acid, and indicates that a carbonyl with a methyl group is to be attached to a compound (Figure 6.25). Similarly, a benzoyl substituent is a carbonyl with an attached phenyl and a formyl group is an aldehyde. Just as having oxygen in an even-sized ring requires there to be another saturated position, the same is true when a carbonyl carbon is a member of an even-sized ring.

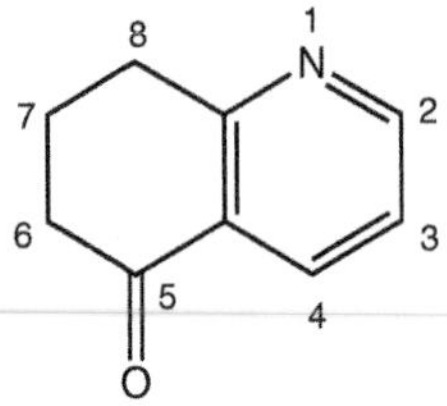

5-Oxocyclohexano[b]pyridine 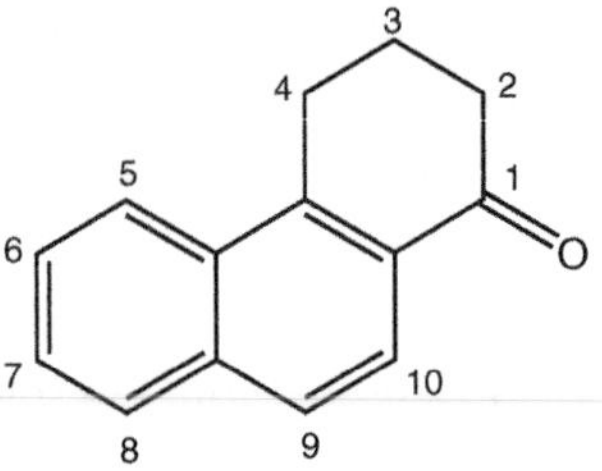   **3,4-Dihydrophenanthren-1(2H)-one**

Acetyl   **Acetic acid**

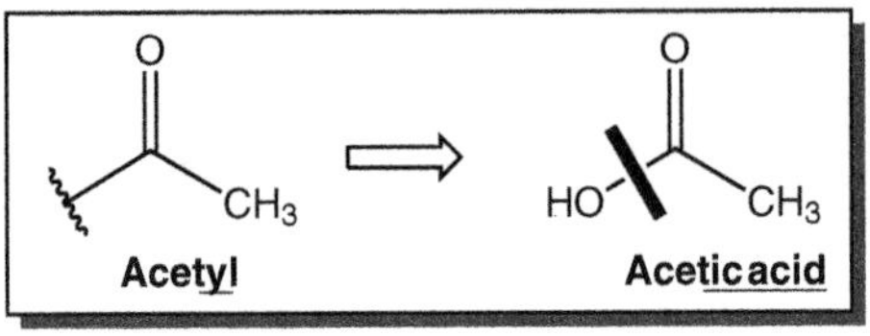

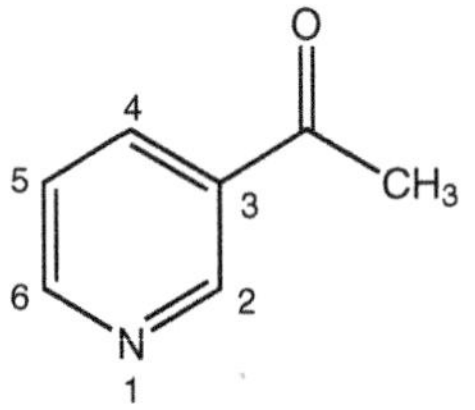

3-Acetylpyridine 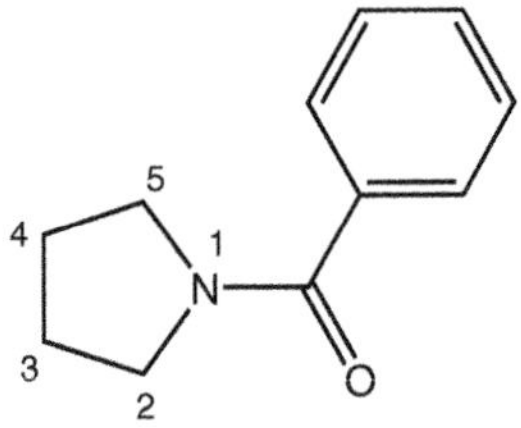   **1-Benzoylpyrrolidine** 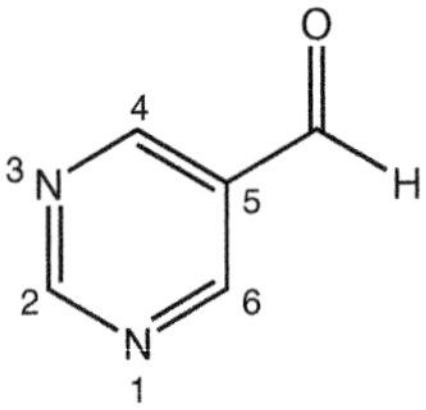   **5-Formylpyrimidine**

Figure 6.25 Nomenclature conventions used for carbonyl groups. In the top two structures, the carbonyl carbon is a member of the ring system while the structures on the bottom have carbonyl-based substituents attached to the ring system by single bonds. The shadow box shows how an acetyl group is derived from acetic acid.

Thus in 3,4-dihydrophenanthren-1(2H)-one (Figure 6.25), the carbonyl occurs at the 1-position and the 2-, 3-, and 4-positions are saturated.

Simple substituents such as methyl, hydroxy, and chloro are self-explanatory. More complex substituents, however, such as hydroxymethyl- and N,N-dimethylaminomethyl-, do require some discussion. When confronted with a complex substituent, it is the right-hand most portion of the name that is attached to the parent compound. The substituent is then built by working backward through the substituent name. This is best understood by examining several examples. We see that in 2-aminomethyl-pyrrole (Figure 6.26), it is a carbon that is attached to the pyrrole 2-position with an amine attached to the carbon. Missing valences on the carbon and nitrogen are filled with hydrogen. Note that the substituent name runs into the parent compound name (pyrrole) with no hyphen. In

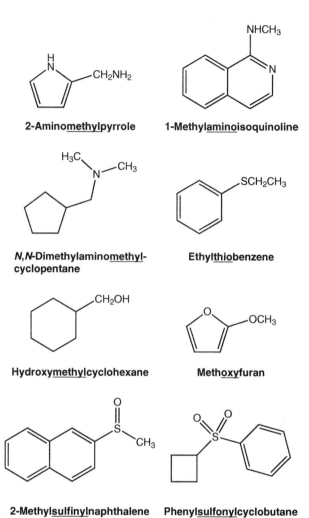

Figure 6.26 A selection of complex substituents is illustrated here. It is always the right-hand most portion of the substituent name (underlined) that is attached to the parent compound.

1-methylaminoisoquinoline, nitrogen is attached to the ring with a methyl group then connected to the nitrogen. For a multicomponent substituent such as in N,N-dimethylaminomethylcyclopentane, the term "methyl" before the cyclopentane indicates attachment of a carbon to the ring followed by a nitrogen and then two methyl groups are added to the nitrogen. The N,N- is somewhat redundant in that it reinforces that the methyl groups are to be attached to nitrogen.

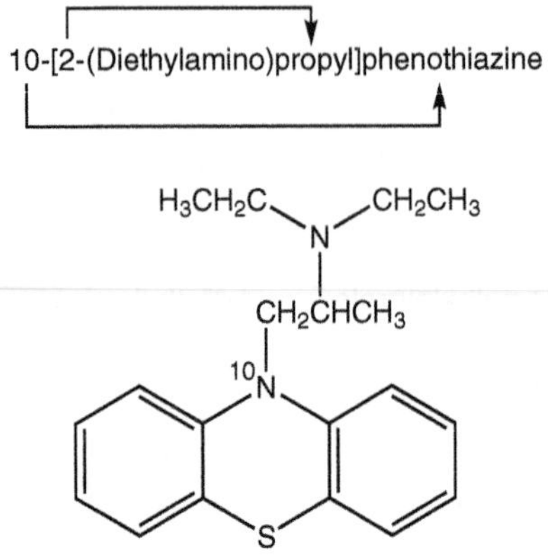

Figure 6.27 More complex substituents are often enclosed in parenthesis or brackets. Numbers outside of the largest parenthesis or bracket refer to positions on the parent compound.

Some substituents are even more complex and parenthesis and brackets are used to maintain the proper order of attachment. Parenthesis typically are used to enclose a relatively simple portion of the substituent while brackets might enclose a larger, more complex portion, including terms in parenthesis, but both symbols are often used interchangeably. One example, 10-[2-(diethylamino)propyl]phenothiazine is shown Figure 6.27. The 10- indicates that everything within the brackets is to be attached to the 10-position of phenothiazine. Thus the bracket term is a complex substituent with a propyl group directly attached to the phenothiazine 10-position. Since no point of attachment is indicated for the propyl group, it is assumed to be attached to the 1-position. Inside the bracket is a 2- and then a term within parenthesis. This indicates that nitrogen (amino) is attached to the 2-position of the propyl group and then two ethyl groups are connected to the nitrogen.

A more complex example is provided by 2-[2-oxo-3-[4-bromo-3-(1-methylethyl)phenyl]propyl]naphthalene (Figure 6.28). Here the 2- indicates that everything within the main bracket (a complex substituent) is to be attached to the 2-position of naphthalene. The last part of the name within the main bracket is propyl and this is attached to naphthalene. At the beginning of that bracket term is 2-oxo-3-[. These numbers refer to the propyl chain which now gets a carbonyl at the 2-position. The substituent within the inside bracket is attached to the 3-position of the propyl chain. Again it is the last term within this bracket (phenyl) that is directly attached to the 3-position of the propyl. At the start of this bracket term, we see that a bromo is attached to the 4-position of the phenyl (where phenyl attaches to propyl is considered

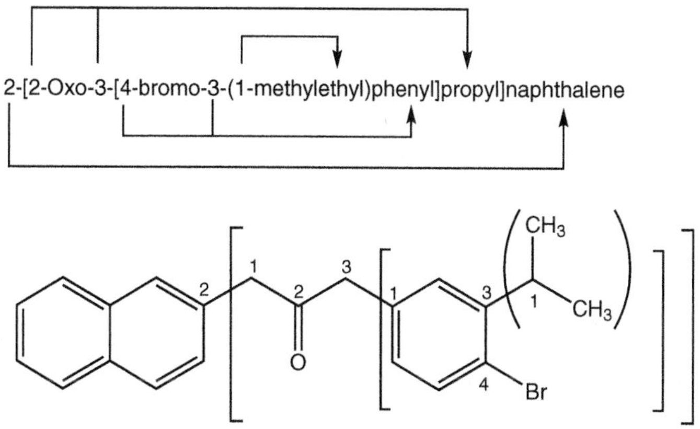

Figure 6.28 An even more complex example involving nested parenthesis and brackets is shown here. Again the number outside of the largest bracket refers to the 2-position of naphthalene, the parent compound. The brackets on the structure correspond to those in the name to help relate the structure to the name.

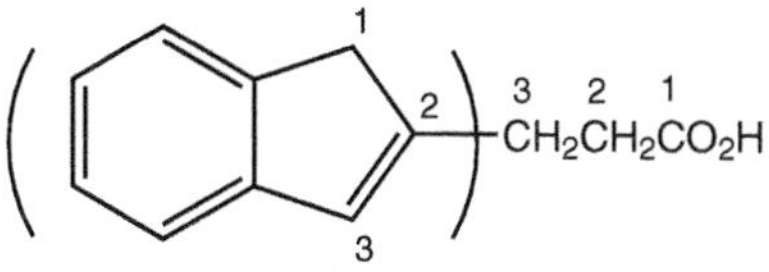

Figure 6.29 When a group that has nonidentical positions is connected to a parent compound, its point of attachment is indicated by a -yl. Here we see the structure of propanoic acid, 3-(1H-inden-2-yl)-.

to be the 1-position of the phenyl ring). The 3-position of the phenyl is attached to the term within the parenthesis, which moving to the end of the name is an ethyl group. The carbon of the ethyl that connects to phenyl is the 1-position, and this also gets a methyl group to complete the molecule.

Substituents with nonequivalent positions require that the point of attachment be specified. This is done with a "#-yl" indicator as in propanoic acid, 3-(1H-inden-2-yl)-. Here the parent compound is propanoic acid and the substituent is 1H-indene, which could be attached to the acid at any of seven different locations. The 2-yl indicates that the attachment is from the 2-position of indene to the 3-position of propanoic acid (Figure 6.29).

Until now all of the substituents have been joined to the parent compound by single bonds. If the attachment is via an exocyclic double bond, it is indicated using "#-ylidene". An example of this is seen (Figure 6.30) in 1-propanamine, 3-(2-chloro-9H-thioxanthen-9-ylidene)-N,N-dimethyl-, (Z)-. The

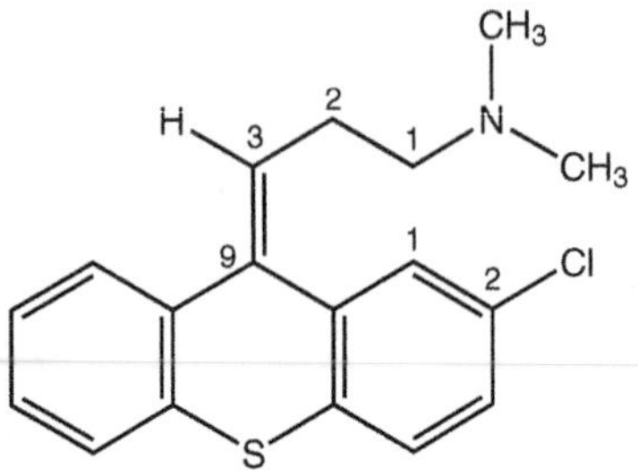

Figure 6.30 A substituent that is attached to a parent compound by a double bond is signaled using -ylidene. The structure shown is for 1-propanamine, 3-(2-chloro-9H-thioxanthen-9-ylidene)-N,N-dimethyl-, (Z)-.

parent compound is 1–propanamine and its 3–position is joined to the 9–position of 9H-thioxanthene by a double bond. Note the use of 9H with thioxanthene even though the compound has an exocyclic double bond at that position. The 9H refers to the particular isomer of thioxanthene employed to construct the compound. In this example, the thioxanthene portion of the molecule has a chloro group at its 2-position and the amine has two methyl groups attached. The final part of the name (Z)- refers to the stereochemistry of the exocyclic double bond. Drawing the molecule with the aminoethyl chain and the chloro-substituted ring of thioxanthene on the same side of the double bond generates a (Z)-double bond.

Occasionally, a molecule will contain multiple copies of a complex substituent. For smaller substituents such as methoxy, di-, tri-, and tetra- are commonly used to indicate two, three, and four of such a group. For more complex substituents however, the prefixes bis-, tris-, and tetrakis- are preferred. An example of this is shown below (Figure 6.31).

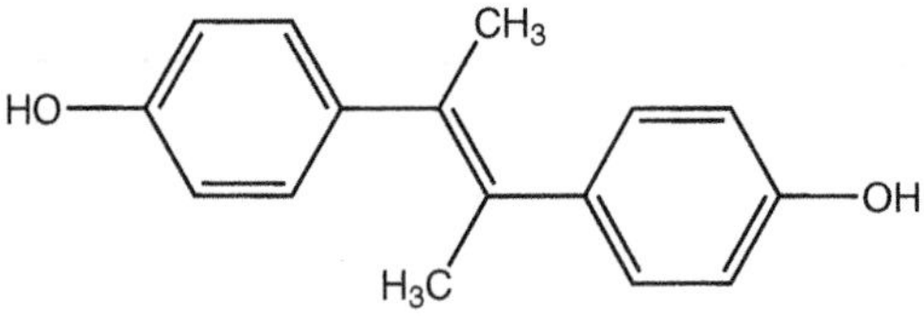

Figure 6.31 The structure of (E)-2,3-bis(4-hydroxyphenyl)-2-butene. Bis- is used to indicated that two 4-hydroxyphenyl groups are to be attached to 2-butene.

Esters often have two part names as in "ethyl acetate". The first part is the alcohol portion, the group that is attached to the oxygen of the ester. Following that is a space and then the acid portion of the name that ends with -ate. This is the group that includes the carbonyl. Thus ethyl acetate is formed from ethanol and acetic acid, $CH_3CH_2O(C=O)CH_3$. A more

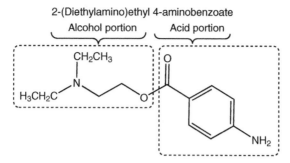

2-(Diethylamino)ethyl 4-aminobenzoate

Figure 6.32 Esters are often named with the alcohol portion listed first (the group attached to the oxygen) and then a space and finally the acid (carbonyl) portion which ends with -ate.

complex ester is found in procaine, 2-(diethylamino)ethyl 4-aminobenzoate. Here the ester is derived from a 4-aminobenzoic acid. The oxygen is attached to an ethyl chain which is substituted at its 2-position with a diethylamino group (Figure 6.32).

When an aliphatic chain is connected to an aromatic ring system, a convention known as *conjunctive nomenclature* is sometimes used. Positions on the aliphatic chain are designated using Greek letters, with the first substitutable position closest to the functional group of highest priority (as assigned for stereochemistry) labeled as "α". If there is no other functional group in the chain, then the position adjacent to the aromatic ring is "α". The next positions in turn are β, γ, δ, etc. Two examples are shown in Figure 6.33. The first is named as a derivative of benzenemethanol. Positions on the ring are specified with numbers, while the aliphatic chain, methanol, has its only carbon labeled as the α-position. The second example has 2–naphthylpropanol as the parent compound. The three carbon chain has a terminal OH group, which has higher priority than the naphthyl ring. Thus the carbon with the OH is the α-position, the middle carbon is β-, and the carbon that connects to the naphthyl ring is γ-.

A salt was defined in Chapter 4 as the product of the reaction between an acid and a base. Most drugs contain functional groups that are either acidic or basic and are converted into salts to impart certain favorable properties such as improved water solubility and crystallinity for ease of handling. Typical basic groups include amines, imines, amidines, and guanidines. These are converted into salts using mineral acids such as hydrochloric, hydrobromic, sulfuric, or phosphoric acid. Alternatively, carboxylic acids such as benzoic, citric, succinic, tartaric, maleic, or fumaric, or sulfonic acids such as benzenesulfonic acid can be employed. Acidic functional groups

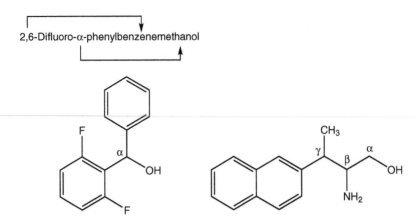

2,6-Difluoro-α-phenylbenzenemethanol

Figure 6.33 One method of differentiating substituents on an aliphatic chain that is attached to an aromatic ring system is conjunctive nomenclature. Positions on the chain are assigned Greek letters with α- as the position adjacent to the substituent of highest priority. This is shown in the example on the left. The right-hand structure, β-amino-γ-methyl-2-naphthylpropanol, provides another example.

such as carboxylic acids, amides, imides, and sulfonamides are often treated with inorganic bases such as sodium or potassium hydroxide for conversion into salts. Figure 6.34 shows the structures of common acids used to form salts from basic drugs. Most commonly salts are formed from an equimolar amount of acid and base and these are designated as 1:1 salts. Dicarboxylic acids such as succinic, tartaric, maleic, and fumaric have the ability to protonate two moles of basic substrate to form 2:1 salts. If a drug molecule has two basic groups of comparable strength, it may also form a 1:2 salt with the acid. The stoichiometry of the salt is sometimes indicated by the appropriate ratio in parenthesis following the systematic name. When drawing salts derived from basic compounds, the most basic site in the molecule gets protonated and a (+) charge is added. The counterion (derived from the acid) is drawn close by the site of protonation with a (−) charge, but is not connected to the parent structure. For salts of acidic compounds, the most acidic proton is removed and a (−) charge is added. The counterion is again drawn close to that site and a (+) charge is added.

Compounds that exist as pure enantiomers are indicated using (*R*) or (*S*) designators in the name. This may be found either before the name as in Figure 6.35 or at the end of the name. When a compound contains more than one chiral center, it is necessary to indicate the configuration at each of these centers. This can be done by specifying the position number of the chiral center followed immediately by *R* or *S* as in the compound (3*R*,6*R*)-6-dimethylamino-4,4-diphenyl-3-heptanol.

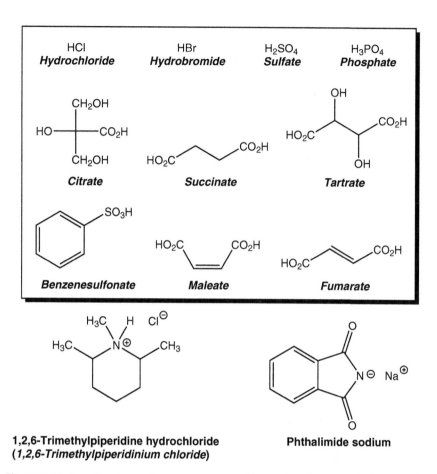

Figure 6.34 *Top*: Structures of some common acids used to form salts (salt names in italics). *Bottom*: Examples of the naming of the salt of a base (left) and an acid (right).

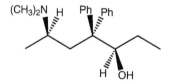

Figure 6.35 Stereochemical designators such as *R*- and *S*- can be located before or after the name. Here, the structure of (3*R*,6*R*)-6-(dimethylamino)-4,4-diphenylheptan-3-ol is shown. Ph is a shorthand method for indicating a phenyl group as a substituent.

An alternative method for specifying configurations at multiple chiral centers is seen in the next example, benzenemethanol, α-(1-aminoethyl)-3-hydroxy-, [R-(R*,S*)] (Figure 6.36). The R outside of the parenthesis gives the absolute configuration of the chiral center of highest priority, i.e.

Benzenemethanol, α-(1-aminoethyl)-3-hydroxy-, [R-(R*,S*)]

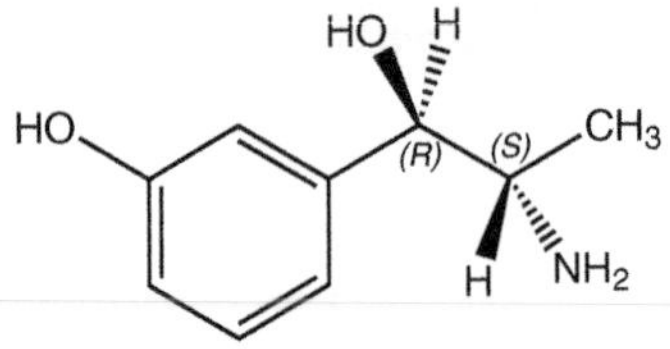

Figure 6.36 The structure of benzenemethanol, α-(1-aminoethyl)-3-hydroxy-, [R-(R*,S*)]. The chiral center with OH has the highest priority and has R-stereochemistry. The other chiral center has the opposite (S*) configuration.

the chiral center to which is attached the atom of highest priority (Cahn–Ingold–Prelog rules). The R^* and S^* within the parenthesis refer to the configuration of each chiral center in the molecule relative to the one of the highest priority. An R^* indicates that a chiral center has the same configuration as indicated outside of the parenthesis, whereas an S^* signifies that a chiral center has a configuration opposite to the highest priority center. Within the parenthesis, the first term is always R^* since this refers again to the highest priority chiral center. As you move from left to right, the priority of the chiral centers decreases. The number of terms inside of the parenthesis is equal to the number of chiral centers in the molecule. In the example above, the highest priority chiral center is the one that has an OH group. This center has the R-configuration. The lower priority chiral center, with the NH_2 group has the opposite configuration, S-. For compounds with two chiral centers, there are four possible combinations:

$$R\text{-}(R^*,R^*) = R,R$$

$$R\text{-}(R^*,S^*) = R,S$$

$$S\text{-}(R^*,R^*) = S,S$$

$$S\text{-}(R^*,S^*) = S,R$$

BRIDGED RINGS

A bridged-ring system is one in which two alicyclic rings have two or more atoms in common. These atoms (*bridge-head atoms*) do not have to be adjacent to one another. Such compounds have a unique numbering pattern and nomenclature.

Table 6.2 Names Used to Indicate the Number of Atoms in a Bridged-Ring System.

No. Atoms	Name	No. Atoms	Name
4	Butane	10	Decane
5	Pentane	11	Undecane
6	Hexane	12	Dodecane
7	Heptane	13	Tridecane
8	Octane	14	Tetradecane
9	Nonane	15	Pentadecane

Bridged-ring nomenclature uses a prefix, suffix, and bracket term as in bicyclo[4.3.2]undecane (Figure 6.37). The prefix indicates the number of bridging rings (bicycle, tricycle, etc.) and whether those rings include any heteroatoms (oxa, thia, aza, etc.).

The suffix term indicates the total number of atoms in the bridged-ring system, excluding any substituents. This is expressed as a hydrocarbon of the appropriate number of atoms even if there are heteroatoms in the ring (Table 6.2). In the example above, undecane indicates that the bridged ring contains 11 atoms.

Inside the bracket are a series of numbers, in decreasing order, separated by periods. In the present example, [4.3.2] indicates that the largest bridge has four atoms between bridge-head positions, the next largest bridge has three atoms, and the smallest is made up of two atoms.

7-Oxabicyclo[2.2.1]heptane   **1,4-Diazabicyclo[2.2.2]octane**   **Bicyclo[1.1.0]butane**

Figure 6.37 Several examples of bridged compounds and their corresponding names.

To draw this ring system (Figure 6.38) begin with a heavy dot to indicate the first bridge-head position. From that dot, draw four atoms and then another dot (the second bridge-head position). Continue from the second dot by drawing three atoms and joining the last of them to the first bridge-head position. Finally, connect the two bridge-head positions by two atoms to construct the last bridge. It can be seen that the resulting compound has bridged seven- and eight-membered rings, as well as a nine-membered *envelope ring*. If necessary, redraw the ring system to impart a sense of three dimensionality, which is especially important if the ring is to have any substituents.

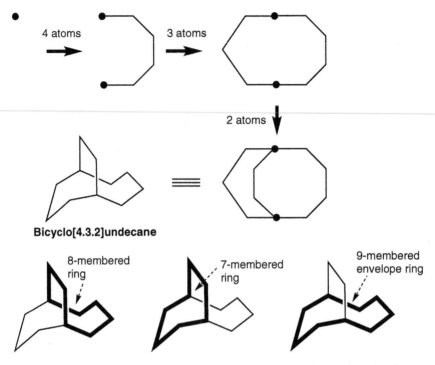

Figure 6.38 The process for drawing the bridged compound bicyclo[4.3.2]undecane. Notice that the resulting compound contains three rings that are seven-, eight-, and nine-membered.

Fused saturated rings are often named using bridge–ring nomenclature, with a zero-atom bridge joining the two bridge-head positions as is bicyclo[1.1.0]butane (Figure 6.37). Occasionally, there are more than two bridging rings in which case there are *main bridges* and *secondary bridges*. The main bridges are the two with the greatest number of atoms, divided as evenly as possible between the main bridge-head atoms. An example is tricyclo[4.3.2.2^{2,5}]tridecane (Figure 6.39). In this example, one notices the superscripted numbers following the 2 inside the bracket. This indicates that a two-atom bridge is to connect the 2- and 5-positions. Thus it is necessary to know how bridged-ring systems are numbered in order to be able to draw this structure.

Numbering begins at a main bridge-head position and continues around the largest bridge to the second main bridge atom. It then continues back to the first bridge atom around the second largest bridge. The next longest bridge is then numbered from the 1-position to the second bridge atom. Finally, any secondary bridges are drawn and numbered from the higher numbered position toward the lower numbered position.

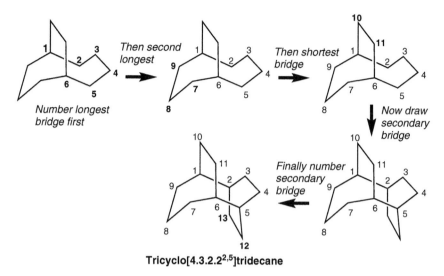

Tricyclo[4.3.2.2²,⁵]tridecane

Figure 6.39 An example of a compound that has a secondary bridge is found in tricyclo[4.3.2.2²,⁵]tridecane.

Heteroatoms should occur at the lowest possible position number and if there are two or more different heteroatoms, then the preference rule $(O > S > N > P)$ is used (Figure 6.40).

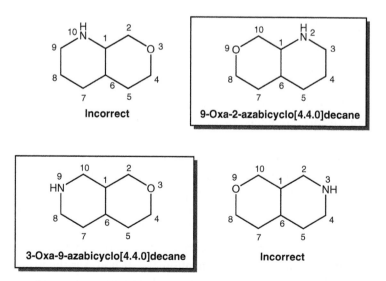

Figure 6.40 If a heterocyclic bridged compound can be numbered in two different ways, use the numbering that puts the heteroatom(s) at the lowest possible position(s). If numbering is still ambiguous, then apply the heteroatom preference rule $(O > S > N > P)$.

In the event that ambiguity remains after employing the rules above, the molecule is numbered so as to assign any unsaturated positions (double bonds, or carbonyls) the lowest possible number (Figure 6.41).

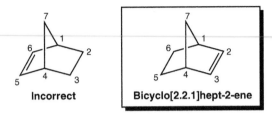

Figure 6.41 If the numbering is still in question after applying the heteroatom rules, then number so as to locate any unsaturated position(s) at the lowest possible number(s).

Finally, if there is still a choice, assign the lowest possible number to any substituents (Figure 6.42).

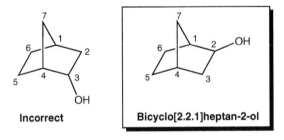

Figure 6.42 The last rule for numbering is to place substituents at the lowest possible numbered positions.

The following compounds (Figure 6.43) are both bicyclo[2.2.1]heptan-2-ols, but are diastereomers. In the compound on the left, the hydroxyl group is oriented closer to the one-atom bridge and is labeled **exo-**, while the other compound has the OH group closer in space to the remaining two-atom bridge (**endo-**). The compounds are named *exo*-bicyclo[2.2.1] heptan-2-ol and *endo*-bicyclo[2.2.1]heptan-2-ol, respectively.

When the two remaining bridges are equivalent, the terms *exo*-and *endo*-do not apply. The structures (Figure 6.44) of bicyclo[2.2.2]octan-2-ol, for example, in which the OH group is, respectively, up or down are nonsuper-imposable mirror-image isomers (enantiomers). If however the bridges are somehow differentiated by having either a substituent or a heteroatom, then the isomer in which the substituent in question is closer in space to the substituted bridge, it is designated as *endo-*.

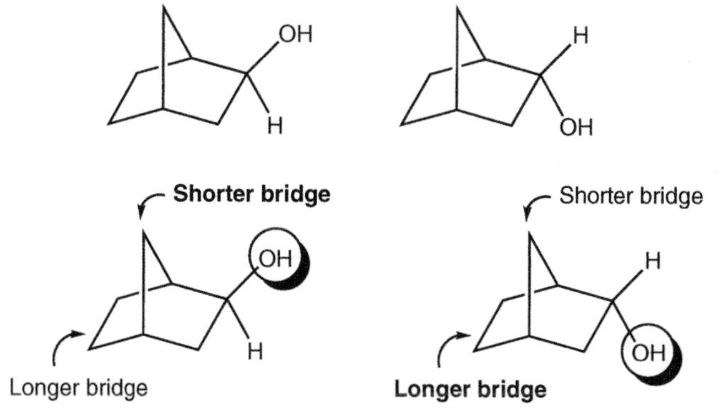

***exo*-Bicyclo[2.2.1]heptan-2-ol** ***endo*-Bicyclo[2.2.1]heptan-2-ol**

Figure 6.43 When a substituent is oriented so that it is closer in space to the shorter of the remaining bridges, it is labeled as *exo-*. If it is closer to the longer remaining bridge, it is *endo-*.

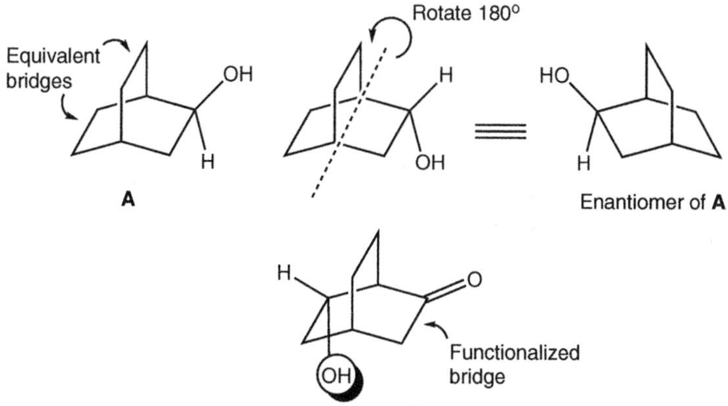

***endo*-6-Hydroxybicyclo[2.2.2]octan-2-one**

Figure 6.44 *Exo-* and *endo-* do not apply when the remaining bridges are the same size unless they are somehow differentiated from one another.

An alternative method to specify the relative orientation of groups in bridged compounds is called *designated nomenclature*. This is best described by examination of its use in naming the anticholinergic compound scopolamine, 9-methyl-3-oxa-9-azatricyclo[3.3.1.0^{2,4}]nonan-7-ol [(1α,2β,4β,5α,7β)]. The bicyclic system is first drawn and numbered as described above, but temporarily omitting any relative stereochemistry. Then the largest ring (the envelope ring) is identified and the highest priority group at the 1-position that is NOT a member of the envelope ring is identified and its orientation is designated as α. The stereochemistry of the highest priority, nonring group at each position is specified within brackets at the end of the name with α-indicating that the orientation is in the same direction as, and β if the orientation is opposite to, the designated group. In scopolamine, 2β and 4β indicate that the hydrogens at these bridging positions (not the oxygen, which is part of the largest ring) are on the opposite face of the ring system as the nitrogen. 5α indicates that the orientation of the highest priority group at that position which is not part of the largest ring (nitrogen) is oriented in the same direction as the designated group (in this case they refer to the same group). The last term, 7β, gives the orientation of the hydroxy group, which is opposite to that of the nitrogen (Figure 6.45).

SPIRO RINGS

Spiro rings are those that share only a single atom. Like bridged compounds, a prefix, bracket, and suffix term are used to name spiro rings. The prefix term includes the term "spiro". If there is more than one spiro atom in a compound, dispiro, trispiro, etc. are used to indicate the number of such junctions. The prefix term also includes any heteroatoms that are part of the spiro ring system, preceded by a number to indicate the position of the heteroatom. The suffix term takes the name of the hydrocarbon containing the total number of atoms in all of the rings, including the spiro atoms. The number of atoms between spiro junctions is indicated within the brackets, separated by periods. These are given in the order in which they occur from one end of the spiro system to the other (Figure 6.46).

To draw a spiro ring system such as spiro[3.4]octane (Figure 6.47), begin again with a dot and then connect the first number of atoms from the bracket; in this case three, and return to the dot. Then continue from the dot the second number of atoms and return to the dot. For dispiro systems such as dispiro[2.1.4.2]undecane, begin as above with two atoms and return to the original dot. Then continue with one atom and draw a second dot. Now draw four atoms and return to the second dot. Finally, connect the second and first dots with two more atoms.

To number a monospiro system, begin at an atom adjacent to the spiro position in the smallest ring and continue around that ring back to the spiro atom which takes the next available number. Then number around the

9-Methyl-3-oxa-9-azatricyclo[3.3.1.0^{2,4}]nonan-7-ol, [(1α,2β,4β,5α,7β)]

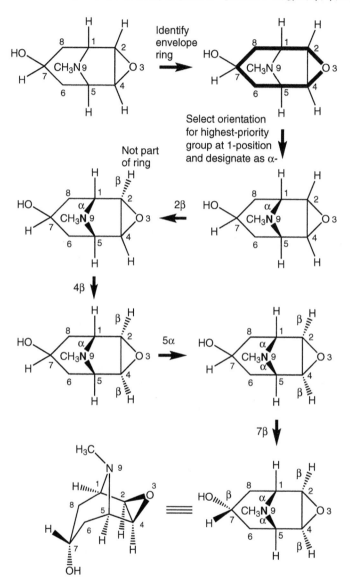

Figure 6.45 Some bridged compounds are named using designated nomenclature. The designated group is the atom of highest priority at the 1-position that is not part of the largest (envelope) ring.

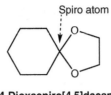

1,4-Dioxaspiro[4.5]decane

Figure 6.46 An example of a spiro ring system.

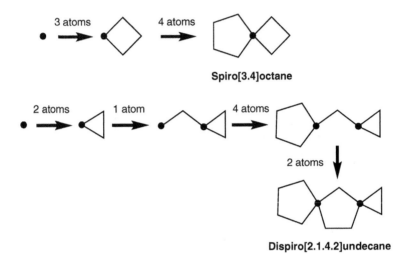

Figure 6.47 *Top*: The process for drawing a ring system with only one spiro atom. *Bottom*: The steps involved in drawing a dispiro ring system.

larger ring (Figure 6.48). The numbering of di- and trispiro rings begins as above around the smallest terminal ring returning to the original spiro position. Then continue so as to give the smallest possible numbers to subsequent spiro positions. Continue numbering around the other terminal ring and finish numbering any remaining positions in the interior ring returning toward the first spiro position.

Additional rules for numbering are the same as for bridged compounds (Figure 6.49). In order:

• Assign the lowest possible number(s) to heteroatoms, employing the preference rule where applicable.
• Assign the lowest possible number(s) to unsaturated positions.
• Assign the lowest possible number(s) to substituents.

It is also possible to have spiro systems comprising certain aromatic ring systems. In such compounds, the term spiro (or dispiro, etc.) is again placed before the bracket. The bracket term however contains the names

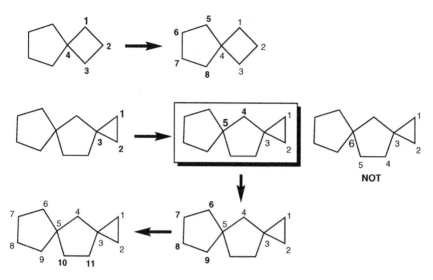

Figure 6.48 Unlike bridged compounds, the numbering of spiro systems begins with the smaller terminal ring. *Top*: Numbering monospiro compounds. *Bottom*: The process of numbering a dispiro ring system.

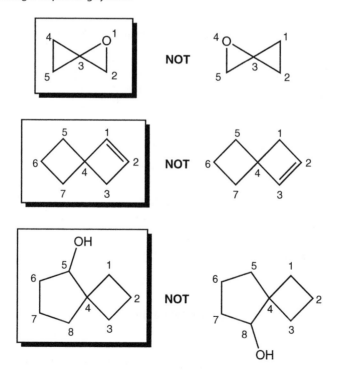

Figure 6.49 If alternative numbering schemes can be applied, then use the rules outlined above for bridged compounds regarding heteroatoms, unsaturation, and substituents to assign a proper numbering to the system.

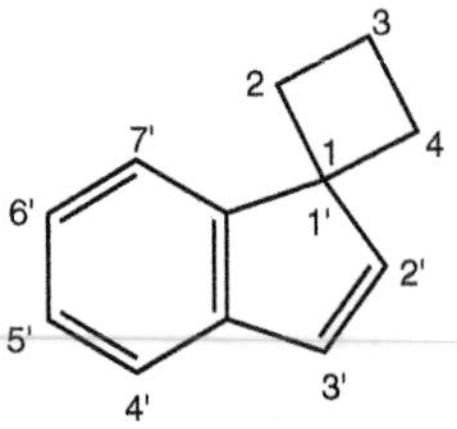

Spiro[cyclobutane-1,1'-1'H-indene] 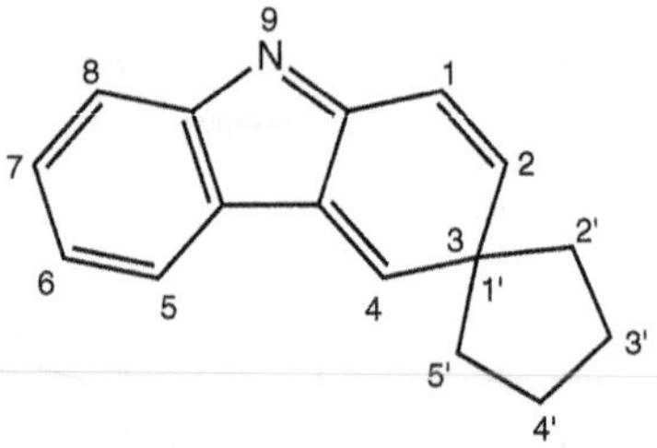   **Spiro[3H-carbazole-3,1'-cyclopentane]**

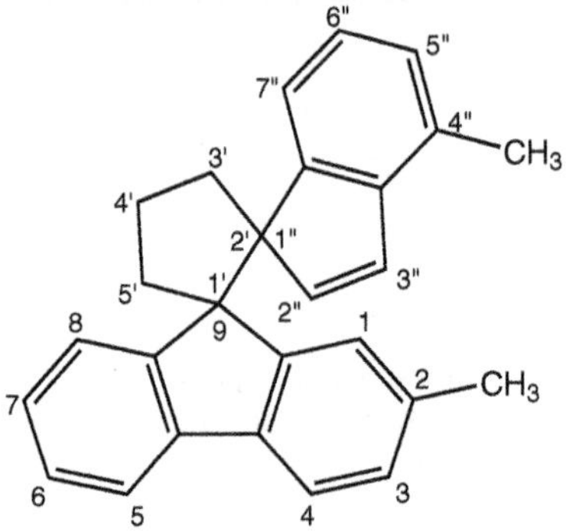

2,4"-Dimethyldispiro[9H-fluorene-9,1'-cyclopentane-2',1"-1"H-indene]

Figure 6.50 When one or more of the rings in a spiro compound is aromatic, the involved rings are named explicitly within brackets, with the point of attachment of each ring specified. Each ring keeps its own numbering, but one ring uses ordinary numbers while the other uses primed, or if necessary, double-primed numbers.

of the ring systems in order of their connection(s). The established numberings of the individual rings is retained, with one ring getting ordinary numbers and the other one(s) getting primed numbers, double-primed numbers, etc. If there is a choice, the numbering that gives the lowest possible numbers to spiro positions is preferred. Spiro positions are indicated by the appropriate numbers from the individual rings. Several examples will serve to highlight the naming and numbering of such systems (Figure 6.50).

SPECIAL CLASSES OF COMPOUNDS

Steroids

Steroids are a class of biologically active natural products that have a common tetracyclic structure. This core structure consists of three fused

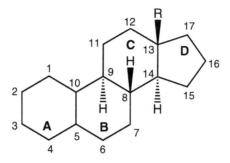

Figure 6.51 Steroids have a characteristic tetracyclic structure with fixed stereochemistry at several positions as shown. The numbering used for steroids is shown on the structure.

six-membered rings and one five-membered ring. The rings are sometimes labeled from left to right as A, B, C, and D, with the D-ring being five membered. Numbering begins at the top of the A-ring and continues in counterclockwise fashion completely around the A- and B-rings. Next the C-ring is numbered clockwise and finally the D-ring is numbered counterclockwise. The stereochemistry of the ring fusions is also generally fixed as shown in Figure 6.51 with the B/C and C/D rings being *trans*-fused. Most steroids possess a methyl group at the 13-position. The A-ring may be saturated, partially unsaturated or aromatic.

Most steroids fall into one of four classes and are named accordingly:
- Cholestanes
- Androstanes
- Pregnanes
- Estranes

Cholestanes have two methyl groups, one each at the 13- and 10-positions. They also have a branched eight-carbon chain attached at the 17-position. The methyl groups and alkyl chain are all oriented above the plane when the steroid is drawn in its standard form shown below. Cholesterol is an example of a cholestane (Figure 6.52).

Androstanes also have the two methyl groups but lack the side chain at the 17-position. Testosterone belongs to the androstane class of steroids (Figure 6.53).

Like the cholestanes and androstanes, *pregnanes* have methyl groups at the 10- and 13-positions. They also have a two-carbon chain at the 17-position in the same orientation as the alkyl chain of the cholestanes. An example of a pregnane is progesterone (Figure 6.54).

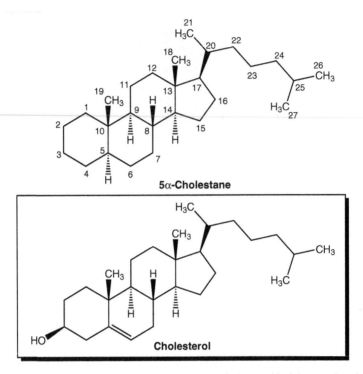

Figure 6.52 5α-Cholestane and its numbering. Cholesterol belongs to this class.

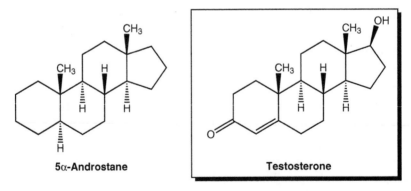

Figure 6.53 The second class of steroids is the androstanes. An example of an andro-stane is testosterone.

The final class of steroids is the *estranes*. Estranes lack a methyl group at the 10–position and a side chain at the 17-position. The A-ring of estranes is often aromatic. Estradiol belongs to this class of steroids (Figure 6.55).

The reader may have noticed in the general structures for the steroid classes that a 5α often precedes the steroid name. In many classes of natural products,

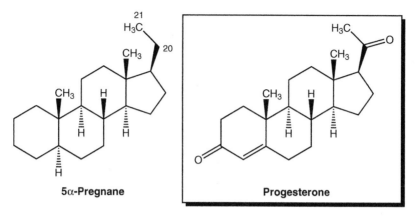

Figure 6.54 The structure of 5α-pregnane. Progesterone belongs to this class of steroids.

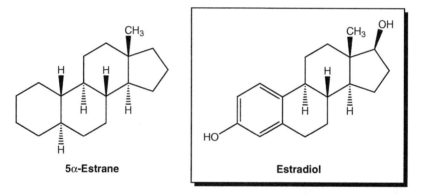

Figure 6.55 Estranes are the fourth group of steroids. The estrogens, including estradiol, are members of this class.

the designators α- and β- are used to indicate, respectively, whether a substituent is behind the plane (down) or in front of the plane (up) of the paper. While the stereochemistry at the 8-, 9-, 10-, 13-, 14-, and 17-positions of steroids is always fixed as shown unless the name specifies otherwise, at the 5-position, it may vary and is therefore specified explicitly. The term 5α- refers to the orientation of the hydrogen at that position. In 5α- steroids, the A/B ring junction is *trans*- and the overall shape of the molecule is relatively planar. As shown in Figure 6.56, 5β-androstane has a *cis*-A/B ring junction, which introduces a severe bend into the overall geometry. In general, most biologically active steroids have either a *trans*-A/B ring junction, or have unsaturation at the 5-position which ensures that the skeleton remains relatively planar.

Substituents are named as usual in steroid names and their orientation is specified using α- and β- along with the position of attachment. Double

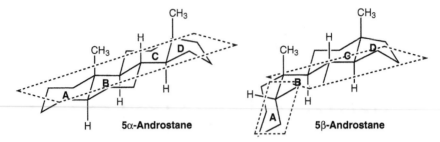

Figure 6.56 Steroids that are saturated at the 5-position must be specified as to the orientation of the hydrogen at that position. For 5α- steroids such as 5α-androstane shown here, the A/B ring fusion is *trans-* and the overall steroid skeleton is relatively planar. A *cis*-A/B junction, as in 5β-androstane, introduces a bend into the overall structure, resulting in the A-ring being nearly perpendicular to the remaining rings.

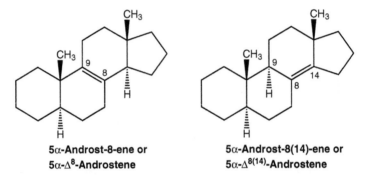

Figure 6.57 Since steroids are generally saturated compounds, any unsaturation must be specified in the name. This is done using either -ene or Δ- with the lower number of the double bond given if it progresses to the next highest numbered position. If not, then both ends of the double bond need to be specified.

bonds within the steroid skeleton are specified with the number of the position from which the bond originates followed by -ene. If the double bond extends to the next highest numbered position, then only the starting position needs to be specified as in 5α-androst-8-ene which implies a double bond between the 8- and 9-positions. If however the double bond is to be drawn to a position other than the next highest number, the end of the double bond is also specified by enclosing its position number in parenthesis as in 5α-androst-8(14)-ene. Double bonds are also indicated occasionally by a capital delta (Δ) with the position(s) of the bond specified by superscripted numbers. An example of this is $5α-Δ^{8(14)}$-androstene (the "ene" ending indicates that the steroid is unsaturated) (Figure 6.57).

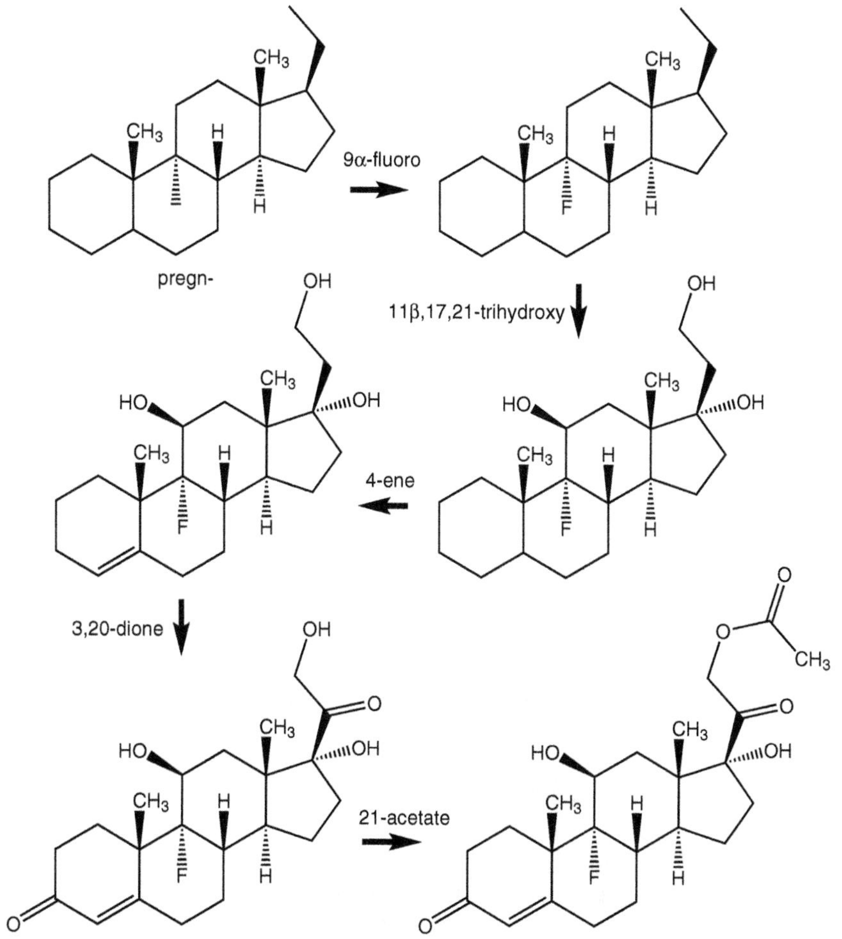

Figure 6.58 The process for drawing a steroid derivative as illustrated for 9α-fluoro-11β,17,21-trihydroxypregn-4-ene-3,20-dione-21-acetate.

Two examples will serve to illustrate how steroids are drawn from their systematic names. The first example (Figure 6.58) is fludrocortisone acetate, which has the systematic name 9α-fluoro-11β,17,21-trihydroxypregn-4-ene-3,20-dione-21-acetate. The process begins by identifying the root name, which in this case is pregn-, indicating that this is a pregnane derivative. This allows for drawing the steroid skeleton with methyl groups at the 10- and 13-positions and a two-carbon chain at the 17-position with the proper stereochemistry at the 8-, 9-, 10-, 13-, 14-, and 17-positions. Next the name is tackled piece by piece, starting with the 9α-fluoro. Thus, a fluoro will take the place of hydrogen that would usually be found at this position, and the fluoro will be down—the same orientation as the hydrogen. The next portion of the name 11β,17,21-trihydroxy indicates that OH groups will be added at the 11-position (up), the 17-position (down), and the 21-position. Since pregnanes have the two-carbon chain up at the 17-position, the additional substituent at this position (the OH) must be oriented down. Following pregn is -4-ene, which tells us to draw a double bond between the 4- and 5-positions. Two carbonyls, one at the 3-position and one on the side chain at the 20-position, are then drawn. Last, we see that the compound is the acetate ester of the 21-hydroxy group.

The second example is ethinyl estradiol which has the systematic name 17α-ethinylestra-1,3,5(10)-triene-3,17β-diol. The root name, estra, informs us to draw a steroid with the appropriate stereochemistry at the 8-, 9-, 13-, and 14-positions with a single methyl group at 13 and no side chain at the 17-position. The term 1,3,5(10)-triene indicates that three double bonds are to be drawn—one between the 1- and 2-positions, another from the 3- to 4-positions, and the last one from the 5- to the 10-position. This renders the A-ring aromatic. Two OH groups are added, one at the 3-position (a phenol) and the other at the 17-position and oriented up (β). Finally, an ethinyl group (acetylene) is also attached to the 17-position, but down (α) (Figure 6.59).

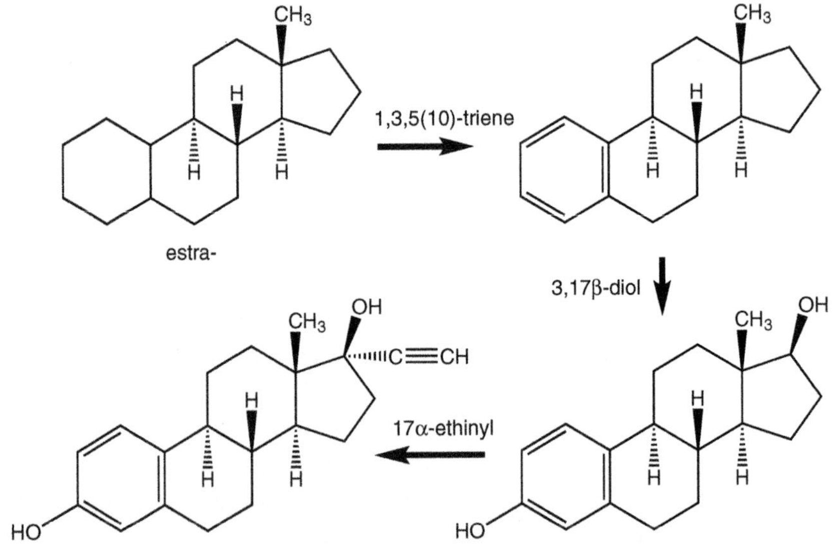

Figure 6.59 A second example of the process of drawing a steroid is shown here using ethinyl estradiol as an example.

Prostaglandins

Physical trauma to the body or exposure to certain microbes stimulates the release of arachidonic acid, which in turn leads to the formation of several prostaglandins that induce a variety of biological responses. Several prostaglandins and their derivatives are employed as drugs for treating gastric ulcers and also for inducing uterine contractions. The naming of these compounds uses a unique shorthand system to avoid the complexity of the true systematic names.

The prostaglandins are derived from prostanoic acid, a twenty-carbon compound (eicosanoid) with the structure and numbering shown below. Numbering begins at the carboxylic acid group and continues around the entire structure in counterclockwise fashion. The stereochemistry, when the prostaglandin is drawn in the orientation shown below, always has the heptyl chain oriented down with an adjacent *trans*-octyl chain on the cyclopentane ring (Figure 6.60).

Prostaglandins are named as $PGX_{\#}$ where X is a letter such as A or D and the subscripted number indicates the number of double bonds in the structure. The letters have significance and refer to the pattern and location of substituents in the cyclopentane ring. Their origin is historical in that when the prostaglandins were first isolated, those found in the ether layer were labeled as PGE derivatives, while those that extracted into a phosphate buffer were labeled as PGF derivatives. In case the reader wonders why these were not called PGP, the isolation was performed in Sweden where phosphate is known as Fosfat. Other prostaglandins were isolated from acid (PGA) and base extracts (PGB). After those, other prostaglandins were simply assigned the next available letters with no special significance. Of the prostaglandins, the most important from a medicinal chemistry viewpoint are the PGE and PGF derivatives. PGE derivatives have a ketone at the

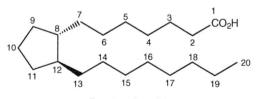

Prostanoic acid

Figure 6.60 Prostaglandins are derivatives of prostanoic acid. Numbering begins at the acid and continues around the entire structure. Note the *trans*-substituents on the cyclopentane ring with the absolute stereochemistry as shown.

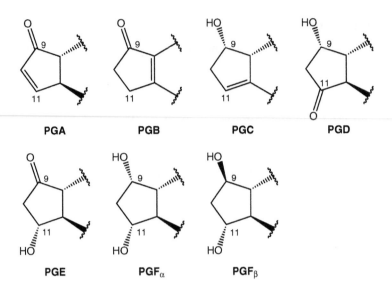

Figure 6.61 The prostaglandins are differentiated by letters indicating the pattern of substitution in the cyclopentane ring.

9-position and an 11α-hydroxy group. Reduction of the ketone gives rise to the PGF analogs. If the resulting 9-hydroxy compound is *cis*- to the 11α-hydroxy group, the compound is a PGF$_\alpha$ derivative, otherwise it is PGF$_\beta$ (Figure 6.61).

When a prostaglandin contains only one double bond, it occurs between the 13- and 14-positions, and has *E*-geometry, as in PGE$_1$. If there are two double bonds, then the second one is *Z*- and extends between the 5- and 6-positions. A third double bond is again *Z*- and located at the 17- and 18-positions (Figure 6.62).

An example using prostaglandin nomenclature is Misoprostol, (*RS*)-16-hydroxy-16-methylPGE$_1$ methyl ester, shown in Figure 6.63. PGE$_1$ is drawn as described above but with two substituents at the 16-position. The stereochemical designator (*RS*) preceding the name refers to the new chiral center formed by the tertiary alcohol and indicates that the compound is a mixture of both stereoisomers at that site (a mixture of diastereomers). Finally, the carboxylic acid is derivatized as a methyl ester to complete the structure of Misoprostol.

Morphinans

The morphinans are a class of natural products that include such compounds as morphine and codeine. These mainly find use as analgesic and

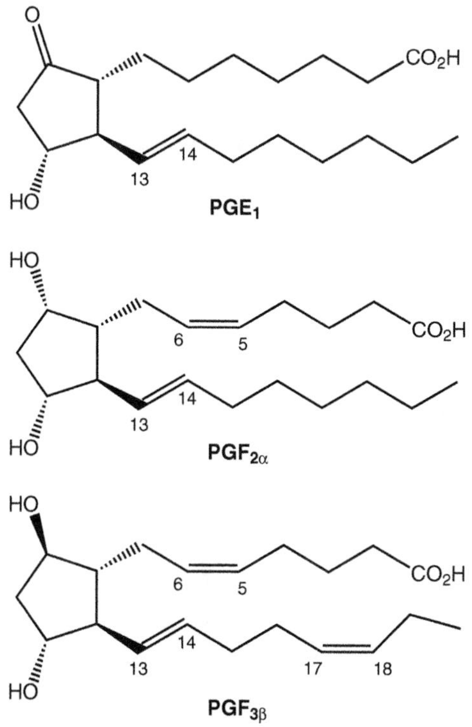

Figure 6.62 Prostaglandins with one, two, or three double bonds are shown here.

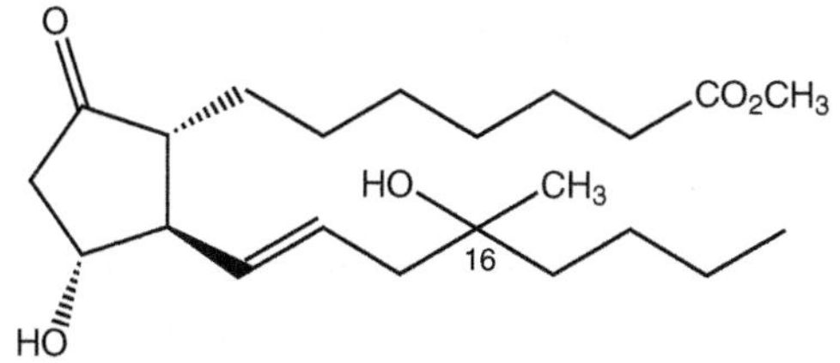

(*RS*)-16-Hydroxy-16-methylPGE₁ methyl ester

Figure 6.63 Prostaglandin derivatives can be named using the prostaglandin short-hand, modified with any substituents.

antitussive agents although they also have several other medicinal applications. Compounds in this class have the basic structure shown in Figure 6.64, and are numbered accordingly. As is evident, the basic structure includes a phenanthrene framework with a bridging piperidine ring. Numbering begins as for phenanthrene through the 10-position, but then the

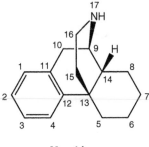

Morphinan

Figure 6.64 The tetracyclic structure and numbering of morphinan is shown here. The stereochemistry at the 9-, 13-, and 14-positions is always fixed unless the name of the compound specifies otherwise.

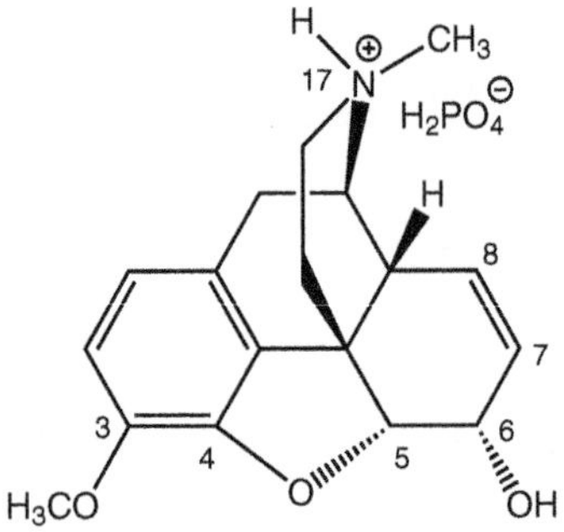

Codeine phosphate

Figure 6.65 The structure of codeine phosphate.

ring junctions and finally, the piperidine ring are numbered with the nitrogen being the 17-position. The parent name morphinan implies this exact structure with the stereochemistry as shown in Figure 6.64.

Several examples will suffice to demonstrate the naming of these compounds. Codeine phosphate (Figure 6.65) is a mild analgesic and a cough suppressant. Its systematic name is morphinan-6-ol, 7,8–didehydro-4,5-epoxy-3-methoxy-17-methyl-, (5α, 6α)-, phosphate (1:1) salt. In the name, we see the parent morphinan listed as well as several substituents. The stereochemical indicators are listed after the name and again we see α- and β- used to indicate the orientation of substituents. Also note that codeine is a 1:1 salt with phosphoric acid. Once the basic morphinan structure is drawn, several familiar substituents such as an alcohol, and a methoxy and a methyl

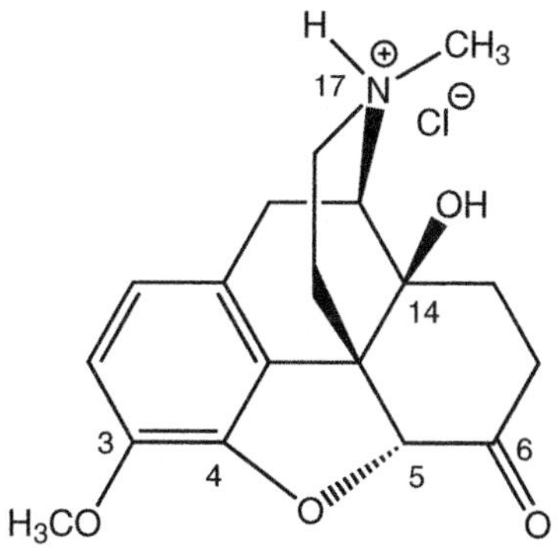

Oxycodone hydrochloride

Figure 6.66 Oxycodone hydrochloride. Note that the OH group at the 14-position takes the same orientation as the hydrogen atom in morphinan.

group can be attached at the appropriate locations. Note that the 6α term refers to the orientation of the alcohol. The term 7,8-didehydro is one that has not been encountered to this point. Morphinan has no double bonds, other than those in the aromatic ring. Terms such as -ene or Δ- have been used to indicate the presence of a double bond in previous examples. Here, didehydro- indicates that we are to remove two hydrogen atoms, one each from the 7- and 8-positions, with the missing valences replaced by a π-bond. Hence this term is simply one more method for indicating that a double bond is to be inserted at the specified location. The term 4,5-epoxy- is used to indicate an oxygen that connects the 4- and 5-positions, with 5α indicating the stereochemistry at that position. The nitrogen is the most basic site in the molecule and is protonated by phosphoric acid, giving it a positive charge. The phosphate counterion is then drawn near, but not connected to the nitrogen.

Oxycodone (Figure 6.66) is another analgesic compound. Its name is morphinan-6-one, 4,5-epoxy-14-hydroxy-3-methoxy-17-methyl-, (5α)-, hydrochloride. In this compound, a carbonyl occurs at the 6-position and there is no 7,8-double bond as in codeine. The 14-hydroxy group, however, occurs at a stereocenter occupied by hydrogen in morphinan itself. When this happens, the substituent assumes the same stereochemical orientation as the group in the parent compound.

The final example is the over-the-counter cough suppressant dextromethorphan, which has the systematic name morphinan, 3-methoxy-17-methyl-, $(9\alpha,13\alpha,14\alpha)$-, hydrobromide. While the name presents no special

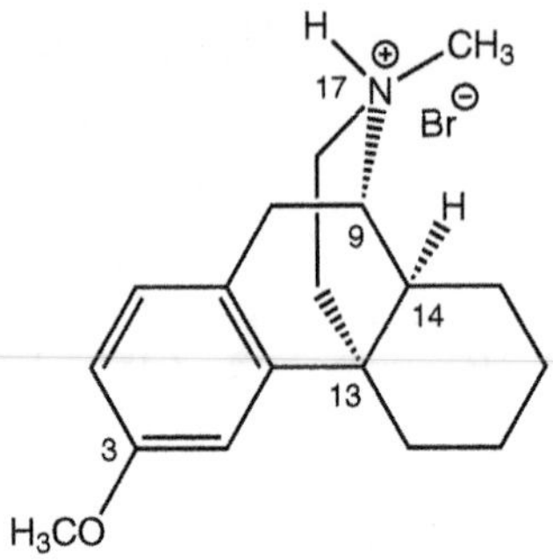

Dextromethorphan hydrobromide

Figure 6.67 Dextromethorphan hydrobromide provides an example of a compound in which the stereochemistry of the three chiral centers of morphinan is inverted.

challenges, the stereochemical designators occur at positions not occupied by substituents. These positions are the three chiral centers of morphinan and the α-designators indicate that each has stereochemistry opposite to that found in morphinan. Thus before attaching any substituents, the tetracyclic framework is the enantiomer of morphinan (Figure 6.67).

CHAPTER 7

Drug Metabolism

When a foreign chemical (a *xenobiotic*) enters the human body it becomes a target for a number of processes that have evolved to transform such compounds into species that can be more easily removed by excretion. These processes are known collectively as *metabolism*. This is a protective measure, the purpose of which is to disarm toxic substances before they cause serious harm, and it generally is very effective. People are inadvertently exposed to many environmental pollutants in the course of daily life through the air they breathe, by absorption through their skin, and by what they eat and drink. Some of these toxins are self-administered, for example, by smoking and drug addiction. Drugs are also taken for legitimate reasons to treat various afflictions and the body handles these in the same manner, as it cannot differentiate between toxic substances and those with beneficial properties. This serves to limit the amount of time that a drug can exert its effect (*duration of action*). A knowledge of drug metabolism is therefore of immense importance to medicinal chemistry. Drugs must be designed so that they are not metabolized too rapidly, lest they have too short a duration of action. It is also important that they do not exert their effects for too long a time period since this may lead to undesired effects. Some drugs rely on the body's metabolic processes to transform an inactive compound into an active drug. These are called *prodrugs* and are used to get around issues of poor absorption or solubility by the active component, or to achieve a slow release of the active drug over time or in a particular compartment of the body.

The principle site in the body at which metabolism occurs is the liver, although other organ sites such as the kidneys, lungs, and gastrointestinal tract have some capacity for metabolism. Metabolism occurs in two distinct phases. In *phase I metabolism* chemical processes either add functional groups to a molecule or interconvert some existing functional groups into others. This generally renders the metabolite more polar than the parent compound and prepares it for elimination from the body. These chemical reactions are generally catalyzed by enzymes and include oxidations, reductions, hydrolyses, and several other processes. It is also important to note that the enzymes that perform these transformations are chiral molecules. If a chiral or a racemic drug undergoes metabolism, diastereomeric drug–enzyme complexes are formed making it likely that the resulting metabolism will be

Organic Chemistry Concepts and Applications for Medicinal Chemistry
http://dx.doi.org/10.1016/B978-0-12-800739-6.00007-3

stereoselective. Because of this, enantiomers may be metabolized by different pathways giving rise to different metabolites. *Phase II metabolism* generally appends a very polar, highly hydrophilic molecule to an appropriately functionalized parent compound or phase I metabolite. This process is called conjugation. Usually, the resulting conjugates lose most or all of their biological activity. Because of their greatly increased hydrophilicity, *conjugates* are excreted primarily in the urine.

In this chapter, the chemistry of phase I and phase II metabolic processes will be explored. As was the case with properties such as acidity and partition coefficients, the types of metabolic transformations that a compound can undergo and the sites at which they occur are determined by its chemical structure and, by extension, the bonds that comprise the molecule.

PHASE I: OXIDATIVE PROCESSES

The largest class of phase I metabolic processes involve oxidative reactions. In its broadest definition *oxidation* involves the removal of electron(s) from a substance such as in the conversion of Fe^{2+} (ferrous) into Fe^{3+} (ferric). To an organic chemist oxidation is either the addition of oxygen to a functional group, or a change in hybridization of a group from sp^3 to sp^2 or from sp^2 to sp by removal of hydrogen. An example of the first type of oxidation is conversion of toluene, $PhCH_3$, into benzyl alcohol, $PhCH_2OH$. Oxidation of a primary alcohol, RCH_2OH, into an aldehyde, $RCHO$, and of an imine, $RCH{=}NH$, into a nitrile, $RC{\equiv}N$, are examples of the second definition of oxidation. Among the more common oxidative metabolic transformations that will be examined in detail are:

- Aromatic hydroxylations
- Epoxidation of alkenes
- Aliphatic hydroxylations
- Oxidation of alcohols to aldehydes or ketones
- Oxidation of aldehydes to carboxylic acids
- Oxidation of sulfides to sulfoxides and sulfoxides to sulfones
- Oxidation of imines to imine oxides
- Oxidation of amines and amides
- Dealkylations on heteroatoms
- Oxidative dehalogenations
- Oxidative desulfurizations

Aromatic rings, whether carbocyclic or heterocyclic, are found in many drugs as well as in a large number of environmental pollutants. One of the major routes of metabolism of such compounds is hydroxylation to produce phenols. Most commonly this occurs on aromatic rings that are electron rich due to the presence of donating groups such as hydroxyl, methoxy, amino, and alkyl. Aromatic rings that have electron-withdrawing groups tend to be more resistant to metabolic hydroxylation. Thus phenolic metabolites are either not observed or are minor metabolites of aromatic rings that contain halides (F, Cl, Br, or I) or nitro, nitrile, carbonyl, sulfinyl, or sulfonyl groups. An example of this is seen with the phenothiazine-based drug, chlorpromazine (Thorazine®) (Figure 7.1). Chlorpromazine has two

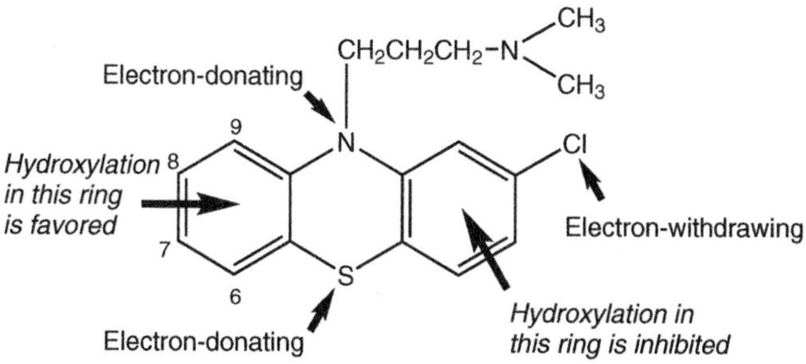

Figure 7.1 The structure of chlorpromazine, showing how existing functional groups affect the site of oxidative metabolism.

aromatic rings. Phenolic metabolites form mainly on the more electron-rich ring encompassing the 6- to 9-positions, while they are inhibited from forming in the chloro–substituted ring. Most often, aromatic hydroxylation occurs at the position that is *para* with respect to the substituent or point of attachment. Electron-releasing substituents can donate by resonance to this position as well as to the two *ortho* positions of the ring. Metabolism at the *ortho* positions, however, is subject to steric crowding that often directs attack to the more accessible *para* position. In chlorpromazine metabolic phenol formation at the 7- and 8-positions is most likely, which are *para* to the nitrogen and sulfur, respectively.

Another drug with two aromatic rings is diazepam, shown in Figure 7.2. The benzene ring that is part of the benzodiazepine system has a chloro substituent at the 7-position and an imine group (part of the ring) attached to the 5a-position. Both groups are electron withdrawing. The lactam nitrogen at the 1-position is a weak donor group that cannot counterbalance the effects of two withdrawing groups. As a result, metabolic hydroxylation at the 6-, 8- and 9-positions is inhibited. The phenyl substituent at the

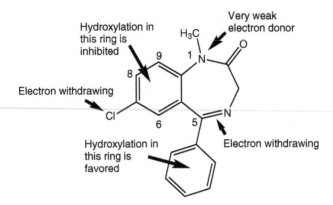

Figure 7.2 The structure of diazepam (Valium®) showing the electronic effects of its various functional groups on aromatic hydroxylation.

5-position is directly connected to an imine, which as mentioned, is electron withdrawing, yet hydroxylation readily occurs at the *para* position of this ring. Why does the imine not inhibit attack on this phenyl ring?

To answer this question we need to understand the three-dimensional structure of diazepam and the mechanisms by which an imine withdraws electrons (Figure 7.3). The nitrogen of an imine exerts a polar effect by virtue of its higher electronegativity relative to carbon. As a result the carbon of the imine has a partial positive charge. Polar effects decrease rapidly, however, with distance and will have little effect at the *para* position of the attached phenyl ring. Electronic effects can be transmitted through much greater distances by resonance and the phenyl ring is in conjugation with the imine group. Resonance requires that the p orbitals of the conjugated system be aligned so that they can overlap. If the phenyl ring was coplanar with the imine then that requirement would be met and the imine would be expected to inhibit oxidative metabolism in the ring. The phenyl ring is not, however, coplanar with the imine. If it were, the *ortho* hydrogens on one side of the ring would be crowded by the hydrogen at the 6-position of diazepam, and the other *ortho* hydrogen would be repelled by the lone pair of the imine nitrogen. To alleviate these steric interactions the phenyl ring is tilted from planarity with the imine to give a lower energy conformation. This has the effect of diminishing both orbital overlap between the π bonds of the phenyl ring and imine and the electron-withdrawing effect of the imine. This conformational preference is general whenever a phenyl ring is attached to another ring by a single bond.

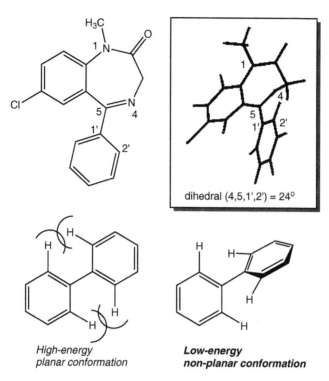

Figure 7.3 *Top*: the two- and three-dimensional structures of diazepam. Note that the phenyl ring is twisted out of planarity with the imine group by 24°. The three-dimensional structure was generated using Chem3D Pro 13.0, Perkin Elmer Informatics, Waltham, MA. *Bottom*: in general, when two aromatic rings are joined by a σ-bond the conformation in which the two rings are not coplanar is lower in energy than the one having both rings in the same plane. Steric interactions between *ortho* groups on the two rings raise the energy of the planar conformation.

When a compound has more than one electron-donating substituent in a ring, aromatic hydroxylation tends to occur predominantly at the position *para* to the one that is the strongest donor. Understanding this preference requires an understanding of the mechanism by which an unsubstituted position on an aromatic ring is metabolized to a phenol. A class of oxidizing enzymes known as mixed function oxidases (MFOs) is responsible for metabolizing aromatic rings. They do this by inserting oxygen into electron-rich π bonds within the rings to form epoxides, which are three-membered cyclic ethers. While epoxidation of ordinary C=C bonds is a common reaction in synthetic organic chemistry, it is extremely difficult to achieve in the laboratory with aromatic π bonds.

Enzymes, however, are able to do this effortlessly. The resulting epoxides are known as *arene oxides*. These species are highly reactive for two main reasons (Figure 7.4):

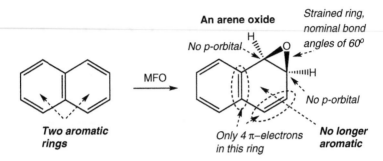

Figure 7.4 Metabolic oxidation of naphthalene forms an intermediate 1,2-arene oxide. This has only one aromatic ring as compared to naphthalene, which has two. The epoxide ring has abnormally small bond angles which produces significant small-angle strain energy. Arene oxides usually undergo further chemical changes that result in opening of the small ring.

- Three-membered rings are highly strained because the bonds that comprise the ring are forced to make bond angles that are significantly smaller than normal. Thus they readily undergo ring opening when attacked by nucleophiles.
- Arene oxides are no longer aromatic because one of the π bonds was used to form the epoxide ring. Thus they no longer benefit from the stabilization energy associated with aromaticity and are instead, high-energy species.

Although arene oxides can occasionally be isolated as metabolites of certain compounds including some polycyclic aromatic hydrocarbons, they most commonly undergo two main types of reactions, which transforms them back into more stable molecules:

- Rearrangement to phenols
- Ring opening by reaction with nucleophiles to form alcohols

Scientists at the National Institutes of Health (NIH) were the first to determine the mechanistic pathway by which arene oxides rearrange to phenols. As a result this is known as the *NIH shift* (Figure 7.5) [1]. The reaction of oxygen with an electron–rich aromatic ring is catalyzed by MFOs to form an arene oxide, which as mentioned is no longer aromatic and possesses considerable small-angle strain. The epoxide can open by breaking

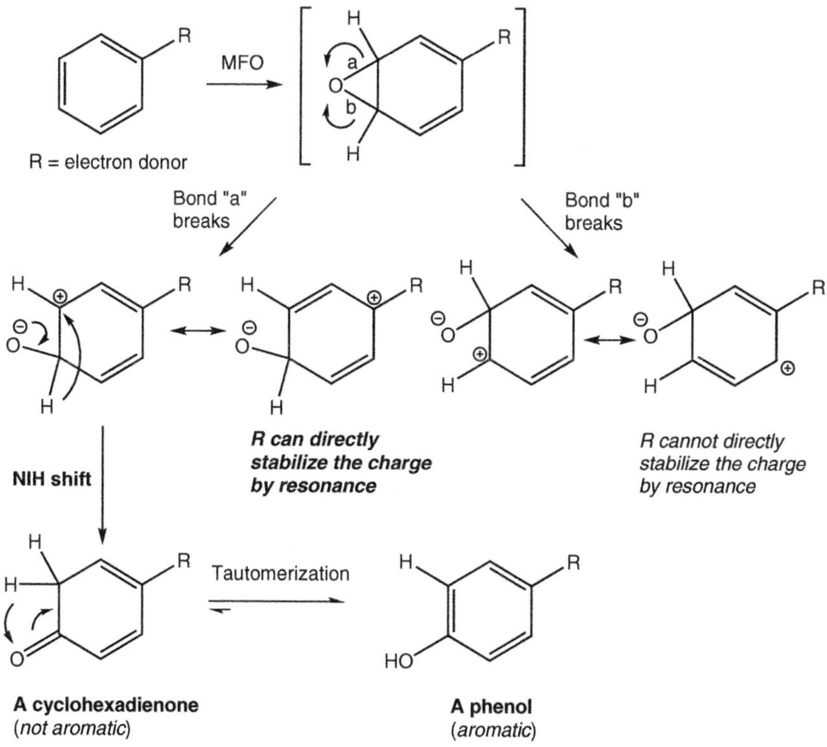

Figure 7.5 The mechanism by which aromatic rings undergo metabolic oxidation to form phenols. An initial arene oxide is formed by the action of mixed-function oxidases (MFO) on electron-rich rings. The oxide can open by breaking either bond "a" or "b". When bond "a" breaks a cation forms on the carbon that is situated meta to the substituent. This can be directly stabilized by resonance from substituent R. This is not true if bond "b" breaks. Thus breaking the bond labeled "a" is the lower energy pathway. The NIH shift then allows migration of a hydrogen from the adjacent position to the cationic center with simultaneous formation of a C–O π bond. The resulting cyclohexadienone, which is not aromatic, then tautomerizes to a phenol which is aromatic.

either of the C–O bonds, labeled as "a" and "b" in Figure 7.5. Both electrons from the bond move to the more electronegative oxygen, leaving either C-3 or C-4 in the ring with a vacant p orbital (positively charged). When bond "a" is broken, the vacant orbital on C-3 can be directly stabilized by resonance from the electron-donating substituent R. If bond "b" breaks, however, the vacant orbital is on C-4 and no direct resonance stabilization from group R is possible. Cleavage of bond "a" therefore leads to a lower energy intermediate and is the preferred route. The actual NIH shift now involves migration of the remaining hydrogen from C-4 to alleviate the charge at C-3. The void resulting at the 4-position when the hydrogen leaves with a pair of electrons is filled by the extra electrons on the oxygen, forming a carbonyl. Finally, tautomerization converts the dienone into a phenol, reestablishing aromaticity in the ring.

The preference for aromatic hydroxylation *para* to the better electron-donating group may now be understood. The arene oxide opens to give rise to the cation that will be stabilized to a greater extent by resonance from the donor group. In the case of phenothiazines such as chlorpromazine, an intermediate 7,8-arene oxide can open in two different directions (Figure 7.6). If it opens so that the oxygen is on the 8-position the resulting cation can be stabilized by resonance from the sulfur. However, sulfides are relatively weak donors because overlap between a 3p orbital on sulfur and a 2p orbital on carbon is required to form the π bond necessary to allow resonance of electron density to occur from sulfur to the cation. Opening of the arene oxide so that the oxygen is attached to the 7-position, however, gives rise to a cation that is readily stabilized by the good electron donor, nitrogen. Thus, we would expect a phenol at the 7-position to be one of the more prevalent metabolites of chlorpromazine, while a phenol at the 8-position would be a relatively minor metabolite [2].

When aromatic rings are oxidized metabolically to arene oxides rearrangement to phenols via the NIH shift is generally the major pathway that is taken because it occurs rapidly. When the arene oxides are relatively stable, however, and have a long half-life then alternative chemistry might occur to give rise to different metabolic products. One such possibility is ring opening of the epoxide by nucleophilic attack of water. This is catalyzed by the enzyme epoxide hydrolase. The enzyme forms a hydrogen bond to one of the lone pairs of the epoxide oxygen. This imparts a partial positive charge on oxygen and weakens the C–O bonds. Since the large enzyme is associated with one face of the epoxide the nucleophile, water in this case, attacks one of the carbons from the opposite side to form a new OH bond as the bond to the epoxide oxygen breaks. Once the enzyme dissociates from the oxygen it too becomes protonated to form another OH bond. The two alcohol groups are on adjacent positions (vicinal) but attached to the opposite faces of the ring. Such compounds are called *trans*-dihydrodiols. This process is illustrated for the polycyclic aromatic hydrocarbon, phenanthrene (Figure 7.7). The 9,10-positions of phenanthrene are susceptible to oxidation to an arene oxide. This epoxide is relatively stable because the compound still possesses two intact aromatic rings. Epoxide hydrolase catalyzes the reaction of this long-lived intermediate to form the *trans*-9,10-dihydrodiol of phenanthrene which is no longer electrophilic and is biologically harmless. The two hydroxyl groups provide a means for attachment of water-solubilizing groups such as glucuronic acid or sulfate—a

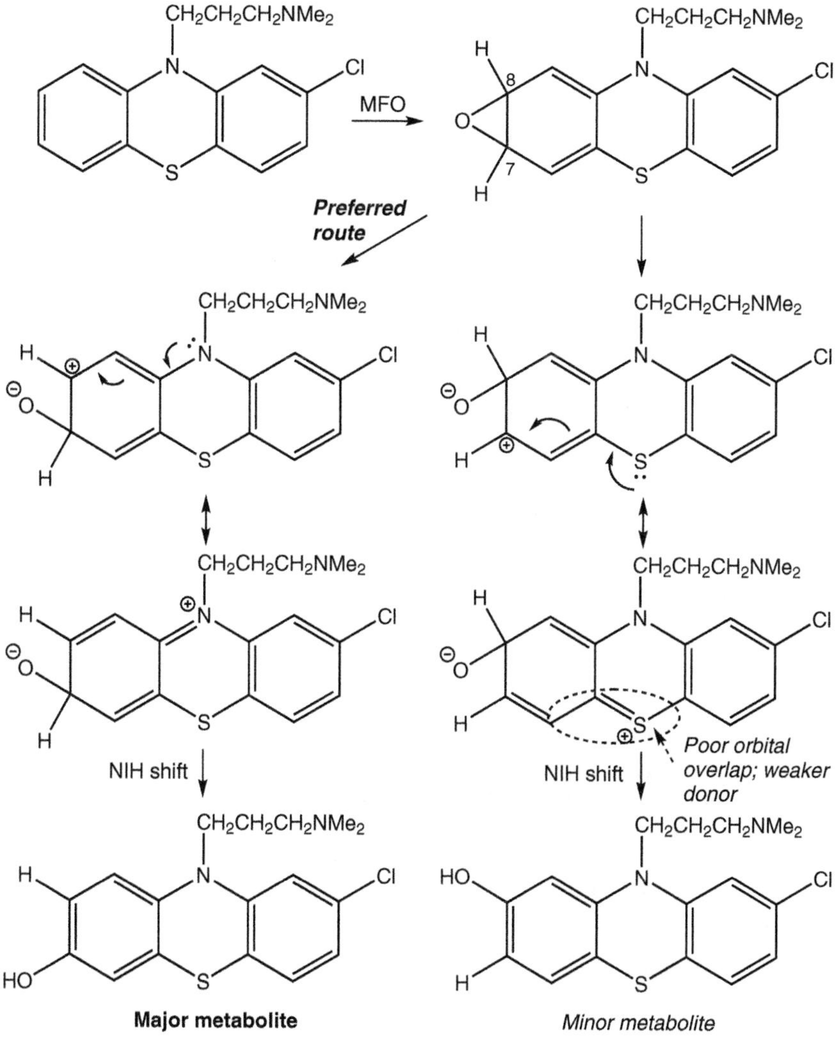

Figure 7.6 The metabolic pathways for conversion of chlorpromazine into phenolic metabolites. The route in which the arene oxide undergoes ring opening to put the cationic center in conjugation with the nitrogen is preferred, leading to 7-hydroxychlorpromazine as a major metabolite [2]. The alternate route requires stabilization of the positive charge by sulfur, which is less favorable due to the relatively poor orbital overlap between sulfur and carbon.

process known as conjugation. This allows for the compound to be more readily excreted in the urine. The process of arene oxide formation and ring opening to form *trans*-dihydrodiols is a major detoxification route for polycyclic aromatic hydrocarbons which are prevalent environmental pollutants.

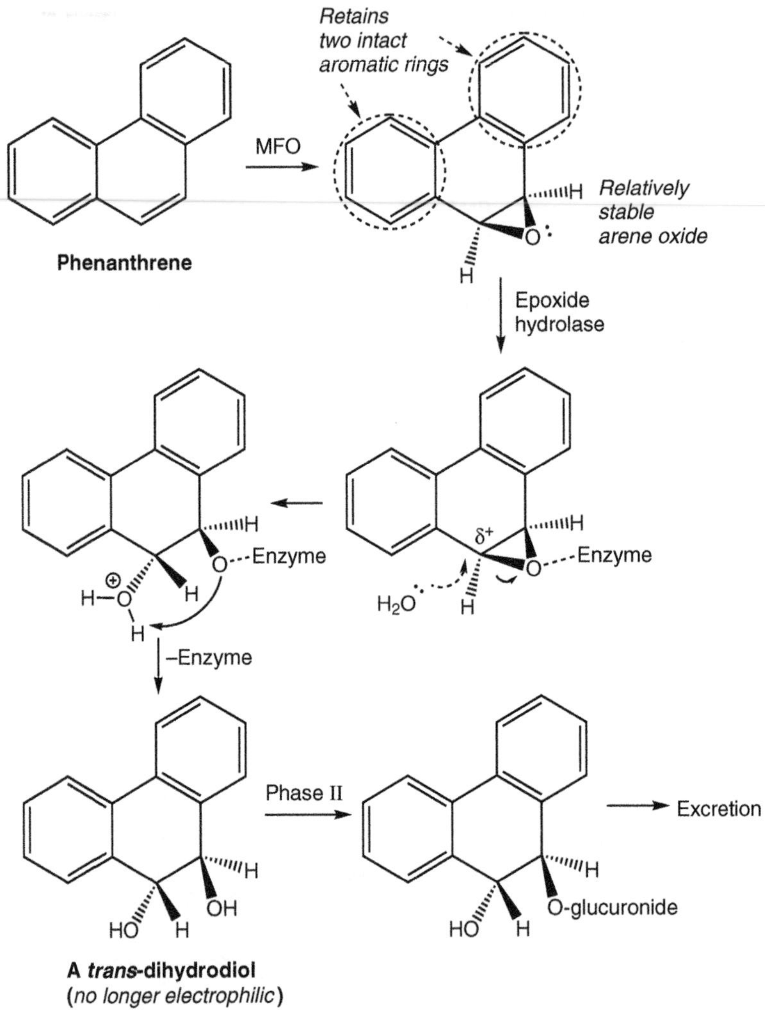

Figure 7.7 Phenanthrene is metabolized to a relatively stable arene oxide at its 9- and 10-positions. The stability arises from the fact that the compound still has two intact aromatic rings. The long half-life of the oxide allows time for alternative chemistry to occur. Here, epoxide hydrolase catalyzes the opening of the epoxide to form a *trans*-dihydrodiol. This is no longer electrophilic and poses little hazard to cells. The hydroxyl groups provide a handle for attachment of glucuronic acid, which dramatically increases the water solubility of the metabolite and allows for its elimination from the body by excretion in the urine.

Water is of course the most prevalent nucleophile in the body, but by no means the only one. Deoxyribonucleic acid (DNA), ribonucleic acid (RNA), proteins, and enzymes also have nucleophilic functional groups and occasionally such groups can compete with water in the opening of arene oxides. When that happens, a covalent bond forms between the compound and the cellular macromolecule, which in the case of proteins and enzymes can modify their usual activity. For nucleic acids a covalently modified base can induce mutations that may initiate tumor formation (Figure 7.8).

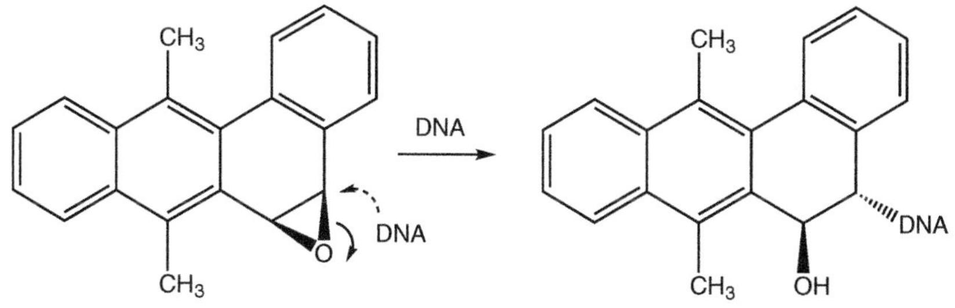

7,12-Dimethylbenz[a]anthracene-5,6-oxide   Covalent DNA adduct that may initiate cancer

Figure 7.8 The environmental carcinogen 7,12-dimethylbenz[a]anthracene is metabolized to its 5,6-arene oxide. This is stabilized because it still has three aromatic rings. In addition, the adjacent methyl group provides steric shielding to the oxide giving it a long half-life. Nucleophilic groups on DNA can open the epoxide resulting in the compound becoming covalently bound to the nucleic acid. This may lead to mutations and ultimately to tumor formation.

When an arene oxide forms on a terminal ring of a polycyclic structure and is converted by epoxide hydrolase into a *trans*-dihydrodiol there still remains one additional double bond in the ring that is no longer part of an aromatic system. This may be subject to further oxidation by MFOs to yet another epoxide. The allylic alcohol of the diol system usually directs the enzyme to attack the double bond from the same face as the OH group. As this is an ordinary epoxide, and not an arene oxide, it cannot rearrange to a phenol via the NIH shift, and has a reasonably long half-life, especially if it is formed in a relatively sheltered area of the molecule known as a bay region. These diol epoxides are electrophilic and may be opened by epoxide hydrolase to form biologically inert tetraols, which get conjugated and eliminated. Alternatively, they may also be opened by reaction with nucleophilic groups on DNA or RNA to form covalently modified bases that lead as mentioned to mutations and tumor initiation. This is one pathway by which the common environmental pollutant benzo[a]pyrene induces tumor formation. For this molecule the *trans*-7,8-diol-*anti*-9,10-epoxide has been determined to be the ultimate carcinogenic metabolite (Figure 7.9). The relative stereochemical designator *anti*- refers to the orientation of the epoxide relative to the benzylic alcohol.

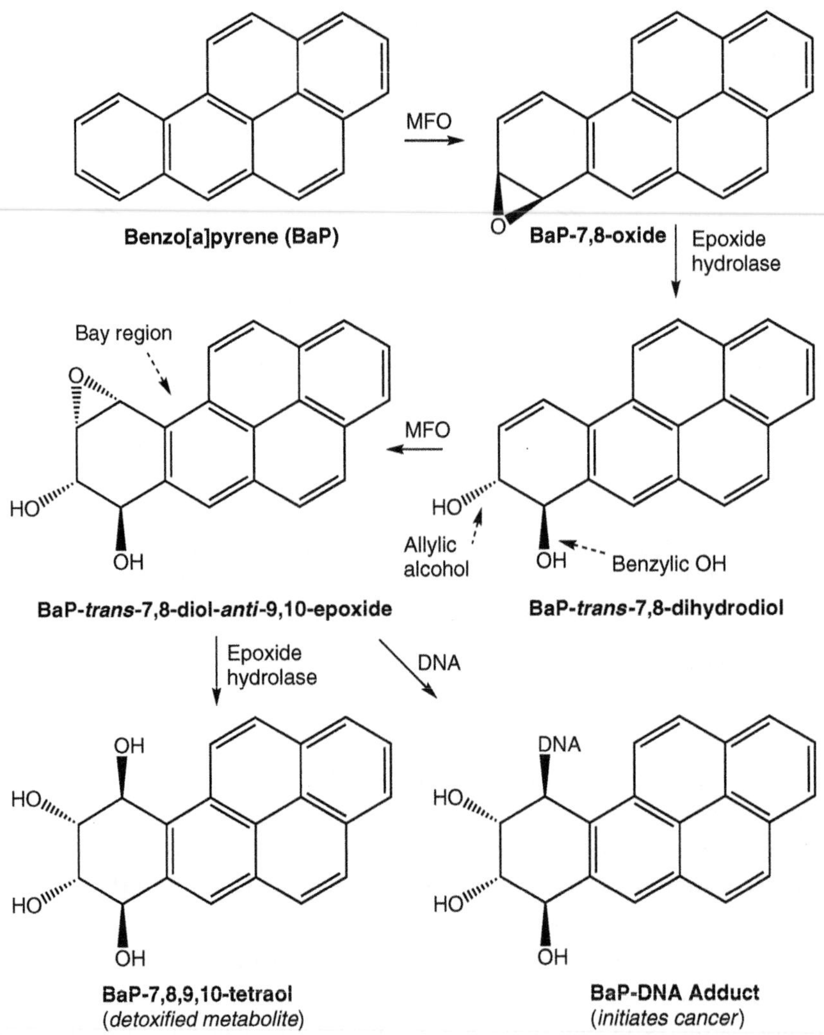

Benzo[a]pyrene (BaP)

MFO

O   **BaP-7,8-oxide** | Epoxide hydrolase

Bay region

MFO

HO\\\\\\

OH   Allylic alcohol

BaP-*trans*-7,8-diol-*anti*-9,10-epoxide

HO\\\\\\

OH ---- Benzylic OH

BaP-*trans*-7,8-dihydrodiol

Epoxide hydrolase

DNA

OH

HO\\\\

HO\\\\\\

OH

BaP-7,8,9,10-tetraol
(*detoxified metabolite*)

DNA

HO\\\\

HO\\\\\\

OH

BaP-DNA Adduct
(*initiates cancer*)

Figure 7.9 Another environmental carcinogen is benzo[a]pyrene (BaP). One site of metabolic attack is the terminal ring which is oxidized to form a 7,8-arene oxide. Epoxide hydrolase opens this to form a *trans*-7,8-dihydrodiol. The remaining C=C bond in the ring may also undergo oxidation to form an epoxide, with the oxidation directed to the same face as the allylic alcohol by coordination of the metabolizing enzyme. This epoxide is now on the opposite face as the benzylic hydroxyl group and is named BaP-*trans*-7,8-diol-*anti*-9,10-epoxide. This is especially stable because it is adjacent to the sheltered bay region of the molecule. It can be opened by epoxide hydrolase to form a detoxified tetraol. Alternatively, DNA can open the epoxide to form a covalent adduct that is responsible for tumor initiation.

Ordinary alkenes are also subject to attack by MFOs to form epoxides. These epoxides, since they are not arene oxides, most commonly undergo hydrolysis catalyzed by epoxide hydrolase to give vicinal diols (Figure 7.10).

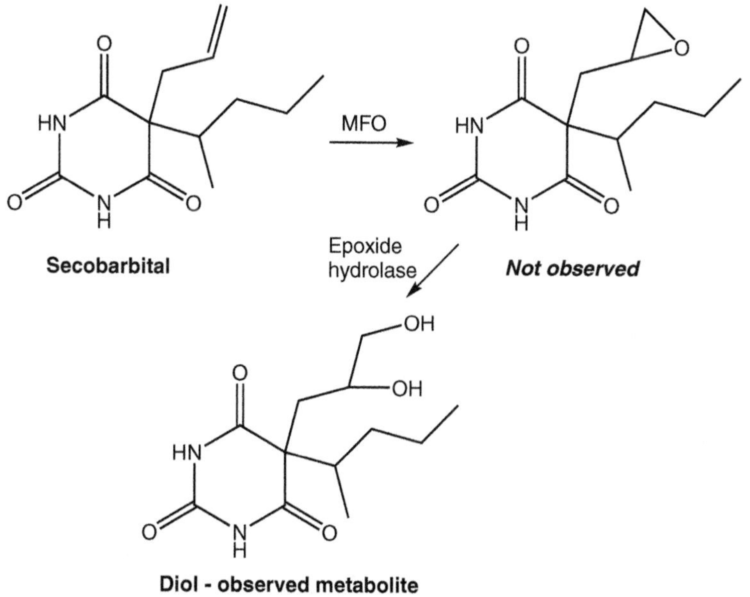

Figure 7.10 Ordinary alkenes also undergo epoxidation mediated by MFO. These have little incentive to rearrange and so they generally serve as substrates for epoxide hydrolase to form vicinal diols. In this example the allylic group of the hypnotic agent secobarbital is epoxidized, but the epoxide is not observed among the isolated metabolites. Instead it is the diol that is found as a major metabolite.

Some aliphatic carbons are also subject to metabolic oxidation. Most usually these are adjacent to a functional group or near the end of a chain of aliphatic carbons. The position immediately next to a functional group is labeled as the α-position and so these are called *α-oxidations* (Figure 7.11).

When oxidation occurs at the last carbon of an aliphatic chain it is termed *ω-oxidation*, and when on the penultimate carbon, it is termed *ω-1 oxidation*. Presumably, these positions are more readily accessible to the enzyme. In each case the product of such an oxidation is an alcohol and the carbon undergoing oxidation must have at least one attached hydrogen atom. These oxidations are also brought about by MFOs, although the mechanism of oxygen transfer is probably different from that used to

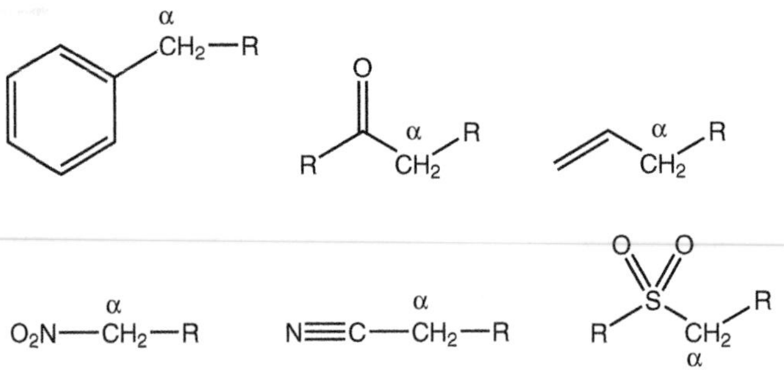

Figure 7.11 The α-position is that which is adjacent to a functional group. Oxidative metabolism often occurs at these positions.

oxygenate aromatic rings and double bonds. Oxygen insertion into an existing C–H bond to give C–O–H is a plausible mechanism. This type of reaction requires an electron-deficient oxygen to be associated with the σ electrons of a C–H bond. As the new C–O and C–H bonds begin to form the existing C–H bond begins to break. Some examples of α-oxidations are shown in Figure 7.12 and Figure 7.13. Figure 7.14 provides examples of ω and ω-1 oxidations.

Alcohols are present in many drug structures and as seen above can also be formed by metabolic oxidation of aliphatic carbons. As long as the carbon having the alcohol group is also attached to at least one hydrogen atom (primary and secondary alcohols) it may be subject to further oxidative metabolism. Thus the enzyme alcohol dehydrogenase catalyzes the oxidation of alcohols to carbonyl compounds. The process essentially requires transfer of a hydride ion from an alcohol mediated by initial deprotonation, to $NADP^+$ which then gets reduced to NADPH as the carbonyl derivative forms. Primary alcohols are oxidized to aldehydes and secondary alcohols are oxidized to ketones. Tertiary alcohols, however, cannot be further oxidized because they have no suitable hydrogen that can leave as a hydride ion (Figure 7.15).

Ketones cannot be oxidized further but aldehydes, being highly reactive, undergo further oxidation to carboxylic acids. This is catalyzed by aldehyde dehydrogenase which essentially inserts oxygen into the C–H bond of the aldehyde. A very common stepwise metabolic process observed with methyl groups, especially those attached to an aromatic ring, is oxidation to a primary alcohol followed by further oxidation to an aldehyde and then to a carboxylic acid (Figure 7.16).

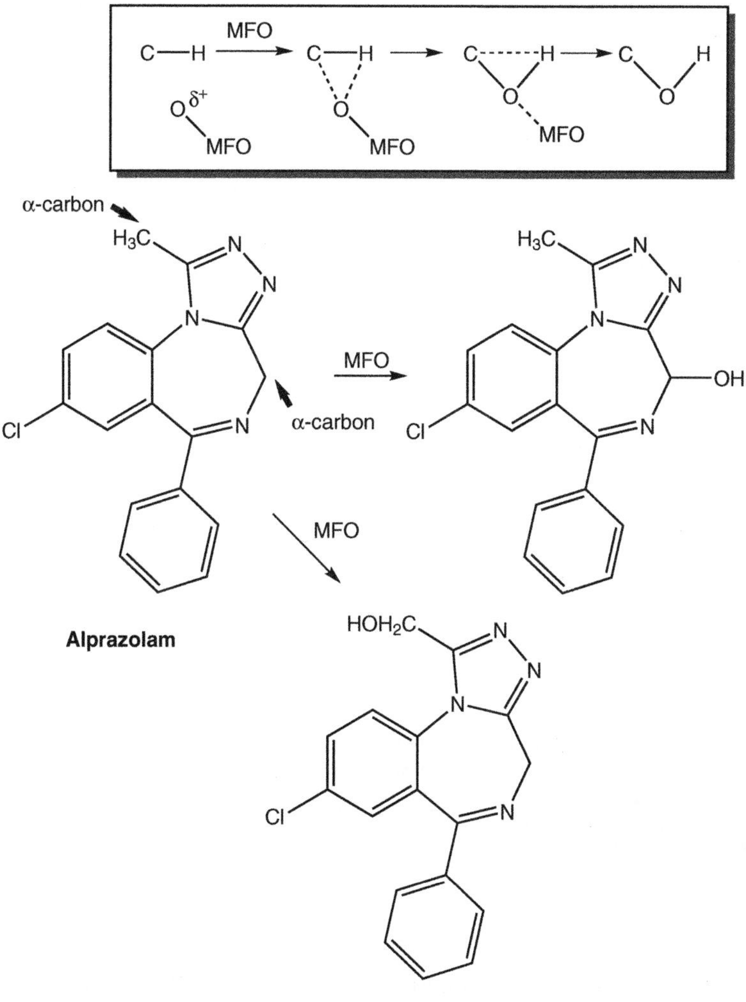

Figure 7.12 *Top*: a plausible mechanism for aliphatic hydroxylation involving insertion of electron-deficient oxygen into a C–H σ bond. *Bottom*: the sedative alprazolam undergoes aliphatic hydroxylation at two α-positions. The methyl group is oxidized to a primary alcohol, while the 4-position, which is α- with respect to the triazole ring and the imine, is oxidized to a secondary alcohol.

Sulfur, as a third row element in the periodic table, has s, p, and d orbitals that can be populated with electrons when making bonds. As a result sulfur can form stable compounds by making two, four, or six bonds. When sulfur forms only two bonds as in sulfides and thiols it is in its lowest oxidation state, while a sulfur that makes six bonds (e.g. sulfones and sulfonic acids) is in its

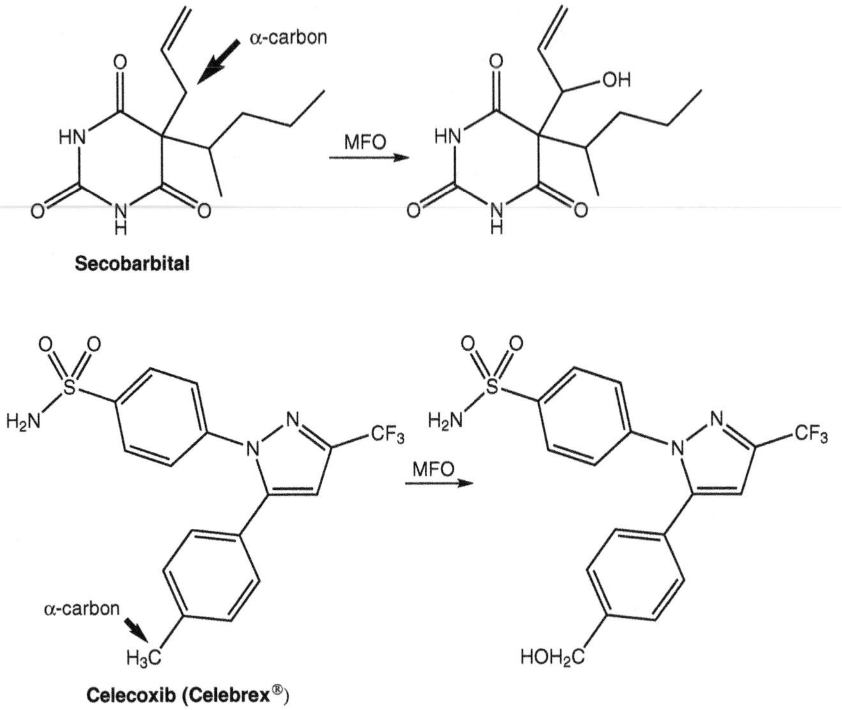

Figure 7.13 Some additional examples of α-oxidations. *Top*: secobarbital is metabolized to a secondary (allylic) alcohol. *Bottom*: a major site of metabolism of the NSAID celecoxib is the aromatic methyl group to give a primary alcohol. This is usually subject to further oxidative metabolism.

highest oxidation state. The MFO system can oxidize sulfur when it is in its two lowest oxidation states. Thus sulfides can be oxidized to sulfoxides, which can be oxidized to sulfones. Sulfones, however, cannot be further oxidized because the sulfur is already in its highest oxidation state (Figure 7.17).

Another substrate for the MFO enzymes is nitrogen, particularly sp²-hybridized nitrogen as found in imines and certain aromatic heterocycles such as pyridine and quinoline. In this case the enzyme facilitates the reaction of the nitrogen lone pair with oxygen giving rise to a species known as an *imine oxide*. Since nitrogen has one more bond than usual it is positively charged, while the oxygen carries a negative charge. The result is a stable compound with opposing charges on adjacent atoms, called a *zwitterion*, which is overall electrically neutral (Figure 7.18).

The mechanism of N-oxidation involves reduction of a flavin unit by NADPH (the flavin monooxygenase system) as depicted in Figure 7.19.

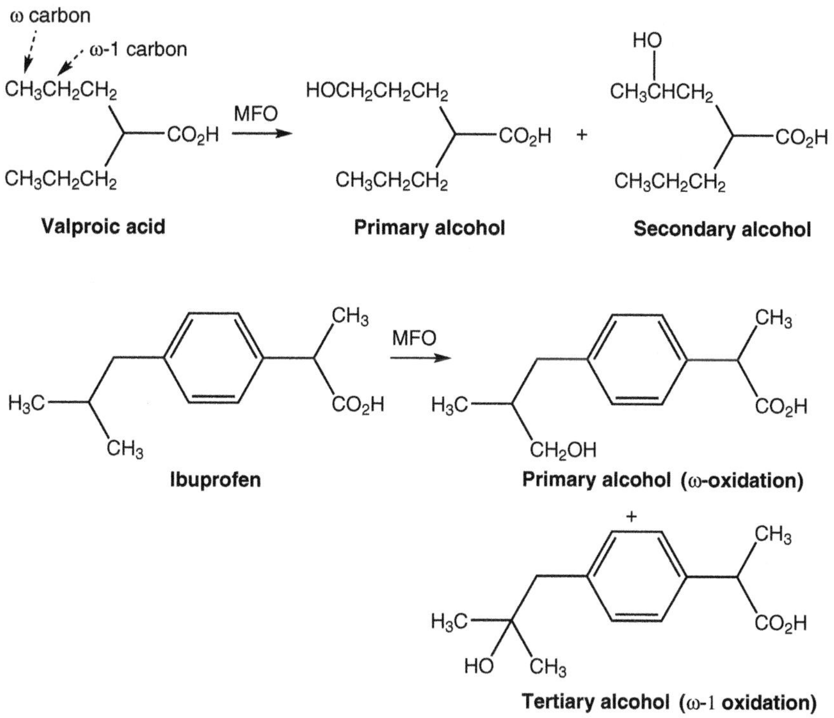

Figure 7.14 When aliphatic hydroxylation occurs on the last carbon of a chain it is called ω-oxidation, while when it occurs on the penultimate carbon it is ω-1 oxidation. Two compounds that are metabolized at these positions are the antiepileptic drug valproic acid and the NSAID ibuprofen.

Hydride anion attacks the sp^2 nitrogen in the middle ring pushing electrons onto the nitrogen to which it was conjugated. As these electrons flow back into the ring they displace another π bond and the electrons are used to form a bond with molecular oxygen to form, after protonation, a hydroperoxide (ROOH). A nucleophile, in this case the nitrogen of an imine, reacts with the electron-deficient terminal oxygen of the hydroperoxide forming the imine oxide after deprotonation. The flavin unit undergoes a ring opening–ring closure sequence and eventually loses a molecule of water to return to the catalytic cycle [3]. Several examples of imine oxidations are shown in Figure 7.20.

Amines and amides can also be oxidized by the pathway described above. Tertiary amines can occasionally be oxidized by this route to zwitterions known as *amine oxides*, which are stable compounds. Primary and secondary amines may also occasionally be substrates for such metabolism, but in those cases the immediate product is a *hydroxylamine*. These usually, but not always,

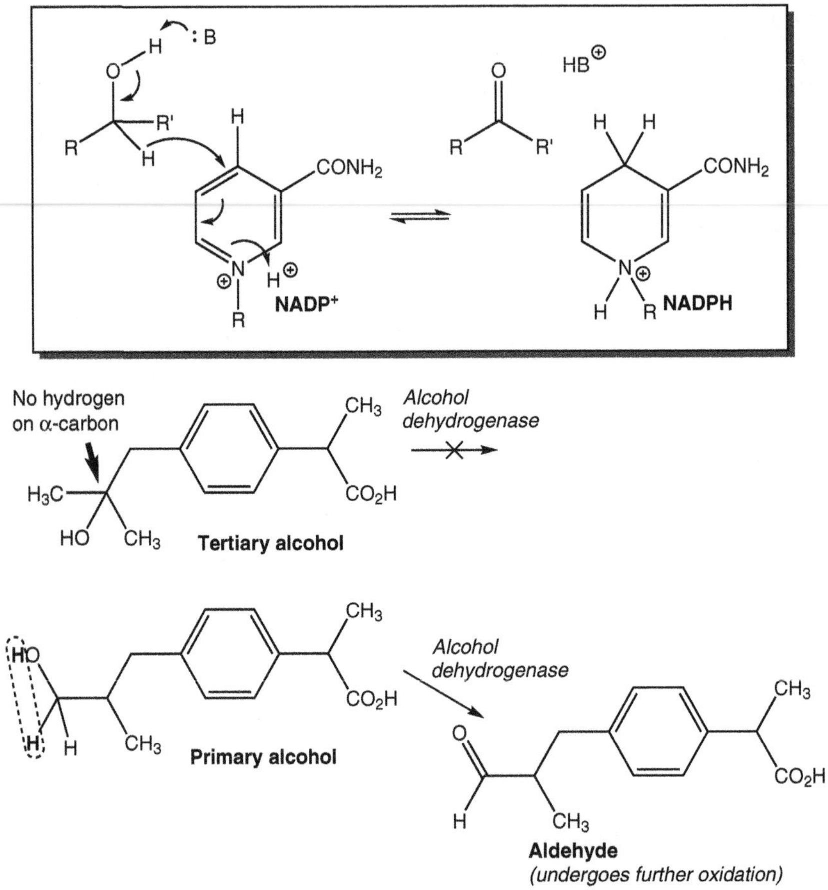

Figure 7.15 *Top*: the mechanism of metabolic oxidation of primary and secondary alcohols to aldehydes and ketones. *Bottom*: the mechanism requires that the carbon with the hydroxyl group is also bonded to at least one hydrogen atom. Tertiary alcohols therefore do not undergo metabolic oxidation by this mechanism. The primary alcohol resulting from ω-oxidation of ibuprofen is oxidized by alcohol dehydrogenase to an aldehyde. This is rapidly metabolized further to a carboxylic acid.

undergo further transformations by elimination of water to form an imine, oxidation of the imine to an oxime, and nucleophilic attack by water to form a reactive carbinolhydroxylamine. Such species undergo hydrolysis to form a carbonyl derivative and hydroxylamine. An example of this is seen with amphetamine, which uses this process to undergo deamination giving rise to phenylacetone as a major metabolite (Figure 7.21).

This is also a pathway taken in the metabolic transformation of primary and secondary aryl amines to toxic intermediates that can sometimes initiate tumor

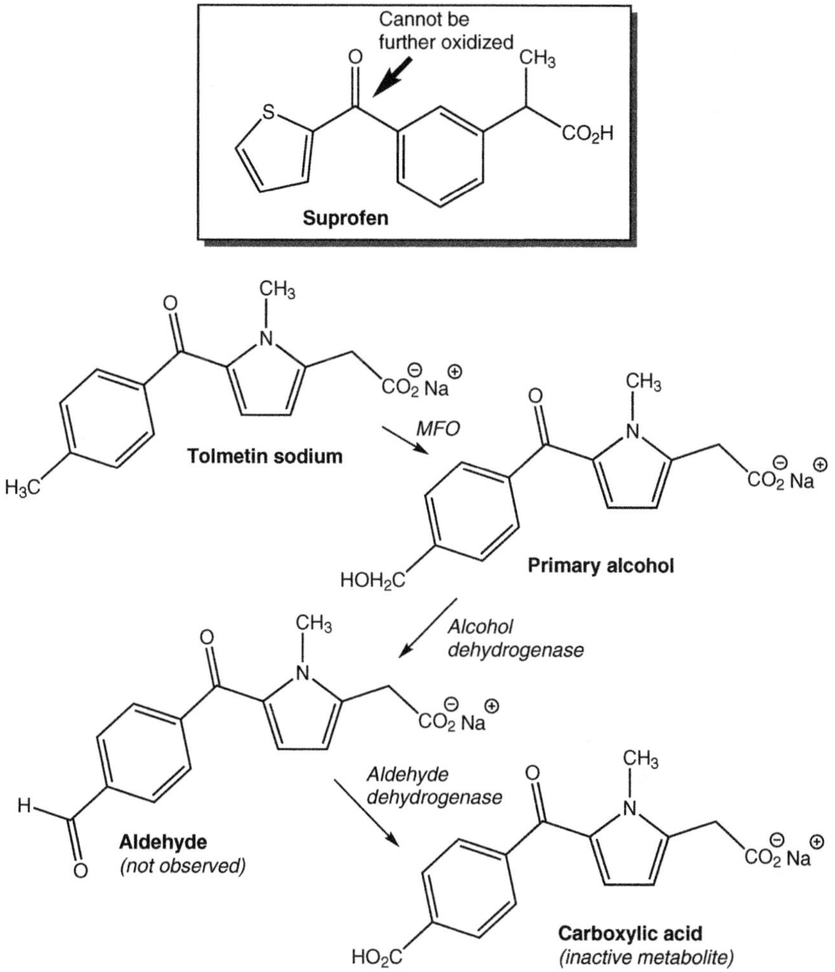

Figure 7.16 *Top*: ketones such as found in suprofen cannot be oxidized further by metabolism. *Bottom*: the NSAID tolmetin is metabolized sequentially on the methyl group, first by the action of MFO to give a primary alcohol, then by alcohol dehydrogenase to an aldehyde, and finally, by aldehyde dehydrogenase to a carboxylic acid.

formation. For primary aryl amines, it is suggested that they first undergo phase II *N*-methylation (see the final section of this chapter) to give a secondary amine (Figure 7.22). This is then transformed by flavin monooxygenase to the corresponding hydroxylamine. Acetylation or formation of a sulfate conjugate (later in this chapter) activates this to a leaving group. Suitably nucleophilic groups from DNA, RNA, or protein attack the aromatic ring at one of the *ortho* positions which pushes electrons from the ring to the nitrogen with

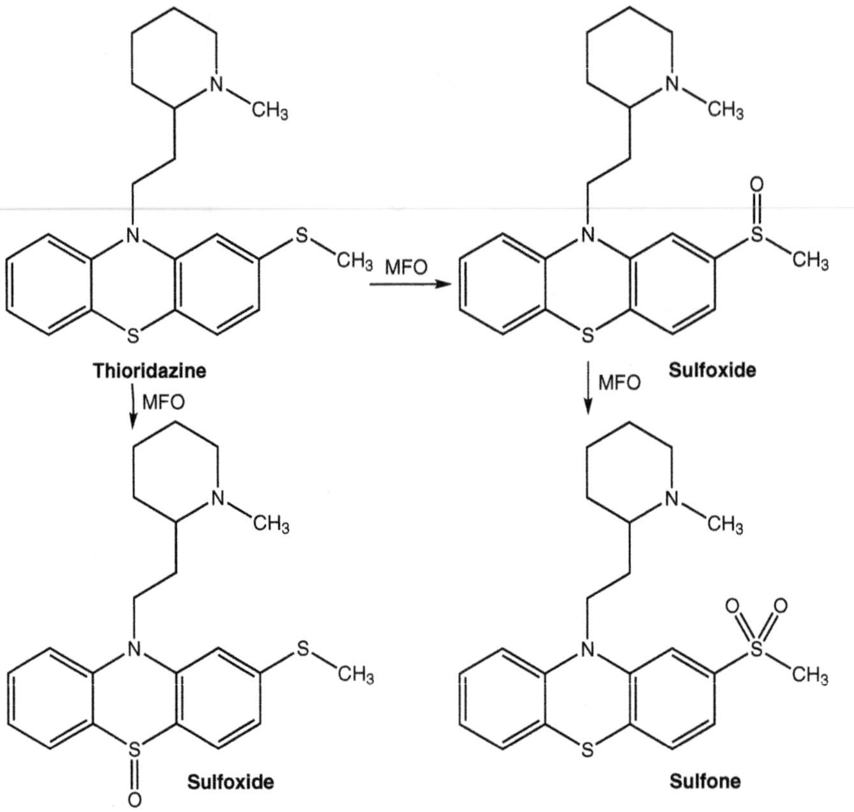

Figure 7.17 The antischizophrenic drug thioridazine is oxidized metabolically on the methylthio group to a sulfoxide and this is further oxidized to a methyl sulfone. Oxidation also occurs on the phenothiazine sulfur to form a sulfoxide.

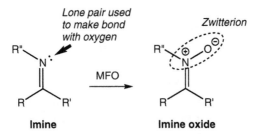

Figure 7.18 The sp^2 nitrogen of imines is subject to metabolic oxidation by the MFO system. The lone pair forms a bond with oxygen to give a zwitterionic species known as an imine oxide.

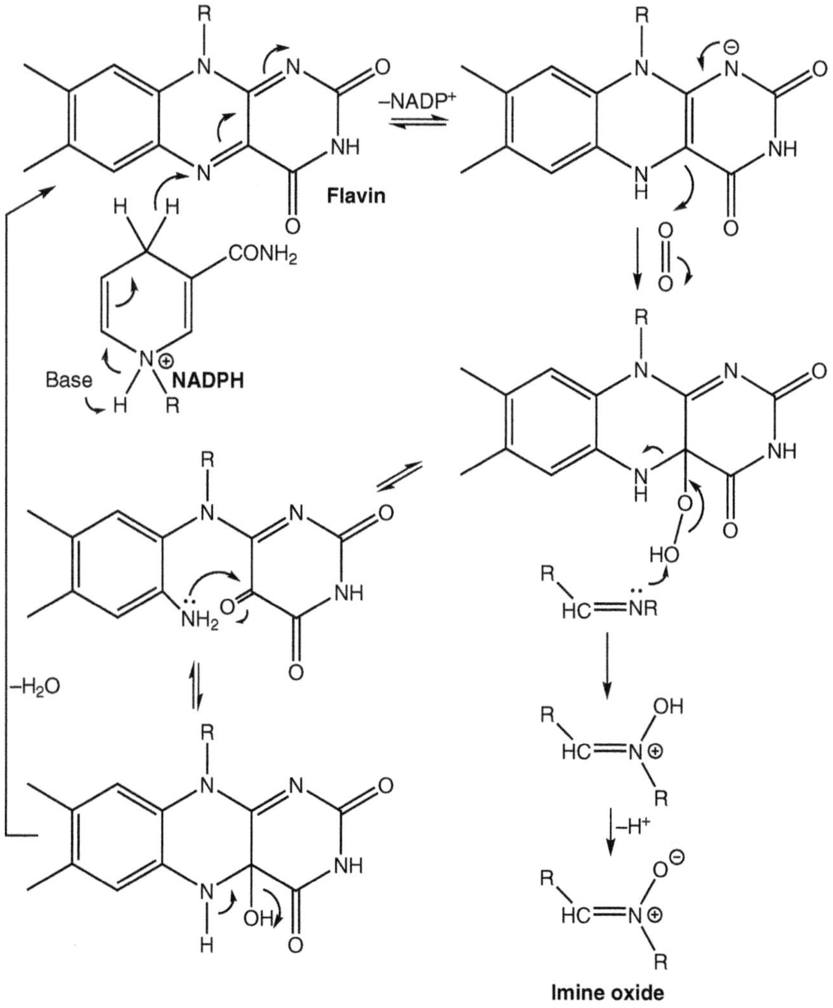

Figure 7.19 The mechanism of N-oxidation relies on a flavin moiety of MFO undergoing reduction by NADPH. The intermediate formed reacts with molecular oxygen to form a hydroperoxide. The terminal oxygen of a hydroperoxide is electrophilic and can react with nucleophiles such as the lone pair of an imine leading to imine oxide formation. The flavin unit undergoes a ring-opening and ring-closure sequence and finally eliminates water to give back the oxidized flavin unit that can continue in the catalytic cycle. Based on Figure 13 in reference [3].

displacement of the leaving group. Tautomerization then restores aromaticity to the ring giving a secondary amine that is now covalently bonded to the cellular macromolecule, leading to altered functionality of the protein and the potential for mutations from covalently modified DNA and RNA [4].

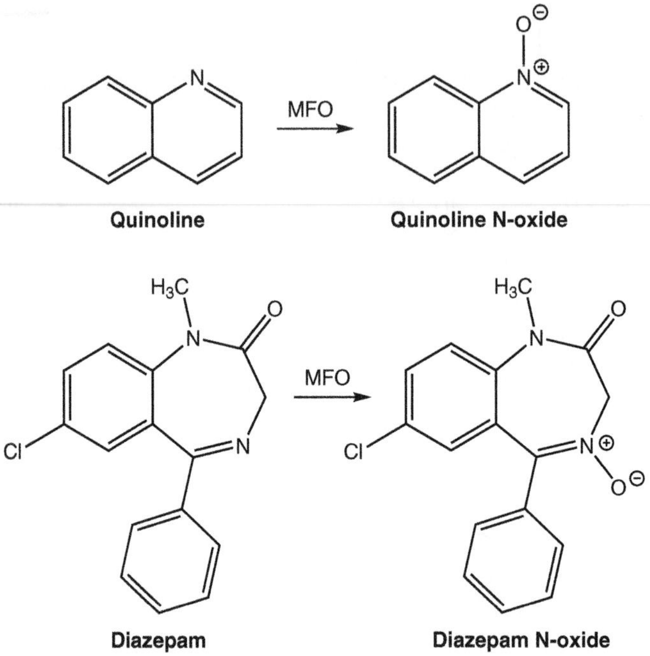

Figure 7.20 Two examples of N-oxidation are shown. *Top*: the sp² nitrogen of aromatic heterocycles such as quinoline undergoes metabolism to quinoline N-oxide. *Bottom*: the imine of diazepam is also metabolized to an N-oxide.

Many drugs contain heteroatom substituents with attached alkyl groups such a dimethylamino, methoxy, or ethylthio. An extremely prevalent metabolic transformation of such groups is the removal of the alkyl group(s) in a process known as *N-dealkylation* (or *O-* or *S-*dealkylation). As with the transformations discussed to this point, this too is an oxidative metabolism. The mechanism for this process will be described for *N*-dealkylation but is identical for *O-* and *S-*dealkylation. A class of flavin–containing enzymes is responsible for *N*-dealkylation, flavin being a compound that oxidizes other compounds, sometimes by removing a single electron via a free radical mechanism. For *N*-dealkylation an electron is removed from the nitrogen lone pair, which initiates a cascade of reactions that ultimately result in cleavage of the C–N bond. Two products always result from *N*-dealkylation—an amine and an aldehyde or ketone. Removal of a primary alkyl group leads to elimination of an aldehyde, while a secondary alkyl group gives rise to a ketone (Figure 7.23). A careful examination of Figure 7.23 reveals that in each case the alkyl group loses one hydrogen in the process

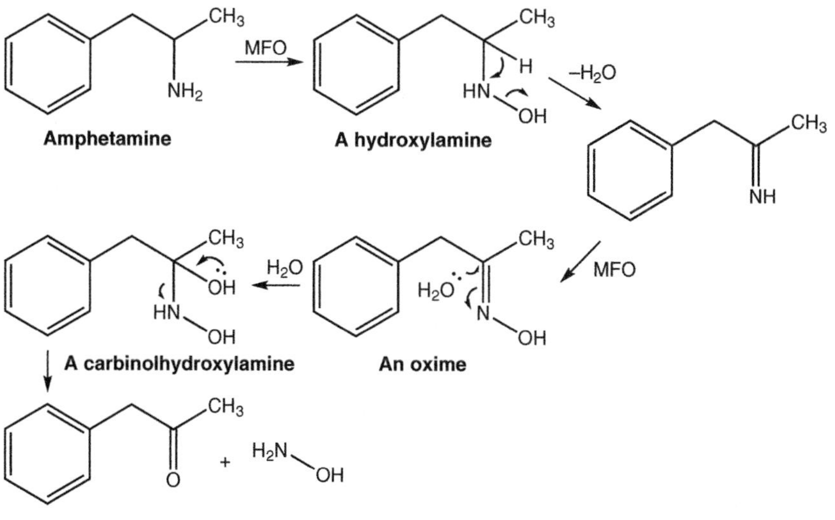

Figure 7.21 When primary or secondary amines undergo N-oxidation the product is a hydroxylamine. This may sometimes provide a pathway for metabolic deamination. Amphetamine is metabolized to a hydroxylamine that then undergoes dehydration to an imine. This is subjected to further metabolic oxidation to give an oxime (possible because the imine had hydrogen attached to the nitrogen). The carbon of the oxime is electrophilic and water attacks to form a carbinolhydroxylamine that is then hydrolyzed to a ketone and hydroxylamine. Thus phenylacetone is observed as a major metabolite of amphetamine.

of becoming transformed into a carbonyl compound. The mechanism of N-dealkylation requires that the alkyl group being removed have at least one hydrogen atom on the carbon. Therefore, tertiary alkyl groups such as *tert*-butyl cannot be removed by this process.

The mechanism of N-dealkylation is shown in Figure 7.24. As mentioned above the flavin portion of MFO removes one electron from the nitrogen lone pair, leaving an unpaired electron (a free radical) on nitrogen. Unpaired electrons are unstable and want to be paired up with a second electron. This electron comes from a neighboring C–H bond. Since a bond cannot exist with only one electron, the remaining electron from the C–H bond leaves with hydrogen (a hydrogen atom) that ends up becoming bound to flavin. Pairing of the one electron from nitrogen and the one from the C–H bond forms new π bond and the resulting species with a positively charged nitrogen is called an *iminium ion*. Iminium ions are electrophiles and readily react with nucleophiles such as water. Water uses one lone pair from the oxygen to react

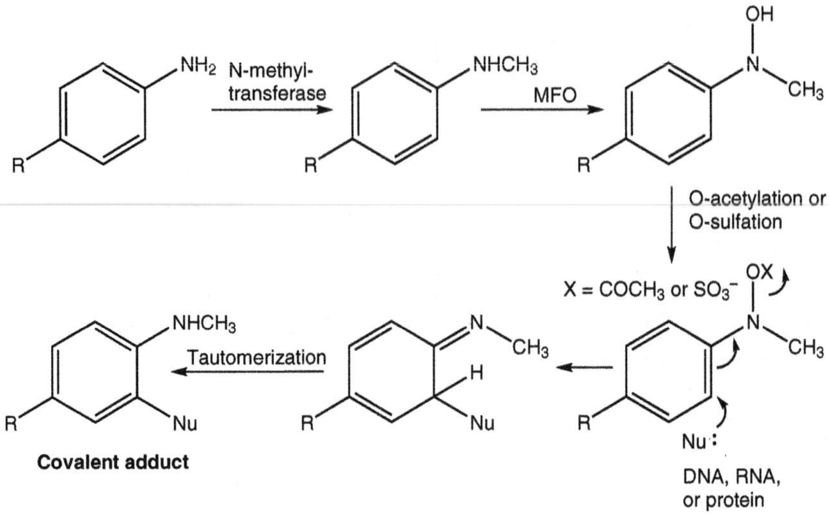

Figure 7.22 N-Oxidation of primary or secondary arylamines again gives rise to hydroxylamines. If the hydroxyl group is activated by phase II conversion to an acetate or sulfate it is converted into a good leaving group. Nucleophiles can attack the aromatic ring at an *ortho* position resulting in elimination of the oxygen group. Tautomerization results in a new aryl amine having the nucleophile covalently bound to the ring. If the nucleophile is a protein or nucleic acid then altered activity or mutations can arise.

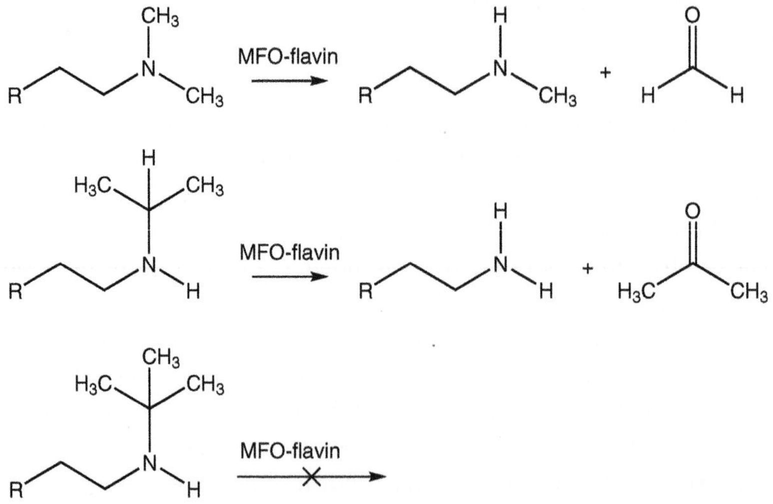

Figure 7.23 Dealkylation of tertiary amines produces a secondary amine and a carbonyl compound, while dealkylation of secondary amines gives rise to primary amines and a carbonyl compound. If the alkyl group is primary it is lost as an aldehyde, if secondary then it is converted into a ketone. Tertiary alkyl groups such as *tert*-butyl cannot be removed by this process.

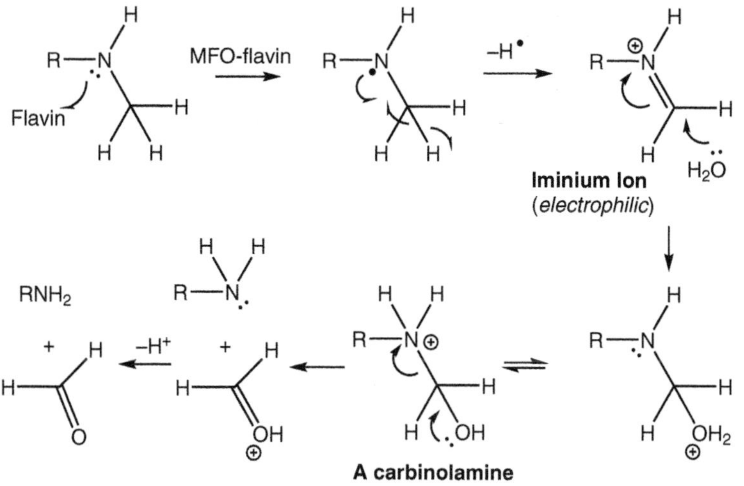

Figure 7.24 The mechanism for N-dealkylation again involves flavin, but this time the flavin removes a single electron from the nitrogen lone pair, leaving behind an amine radical. Note that it is conventional to use curly arrows with a single barb (fishhook) to designate the movement of a single electron, while curly arrows with a double barb are used to indicate movement of a pair of electrons. Unpaired electrons prefer to be paired so a neighboring C–H bond breaks homolytically, with one electron pairing with the lone electron on nitrogen to form a π bond and the remaining electron leaving with hydrogen. Since nitrogen now has four bonds it is positively charged and this highly electrophilic species is called an iminium ion. Water can attack the carbon to form a carbinolamine that then hydrolyzes to produce an amine and, in this case, an aldehyde.

with the carbon of the iminium ion. This forces a pair of electrons from the C=N π bond back onto nitrogen, reestablishing a lone pair. The resulting species has a protonated OH group and an amine both attached to a carbon. The amine is basic and the protonated OH group is acidic and so a proton is transferred from the OH_2 group to the amine to give a new species with an ammonium ion and an alcohol attached to a single carbon (a *carbinolamine*). At this point a lone pair from oxygen moves into the carbon and the electrons in the C–N bond are forced completely onto the nitrogen to give an amine and a protonated carbonyl compound. Loss of a proton from oxygen to water or any other basic substance in the vicinity completes formation of the carbonyl compound (in this case formaldehyde).

When the drug contains a tertiary amine N-dealkylation can occur either once to give a secondary amine and a carbonyl compound or twice, to give a primary amine and two carbonyl compounds. If the carbon on the other side of the nitrogen also contains at least one hydrogen atom, then the C–N bond can also be cleaved to give an aldehyde or ketone and ammonia. This is called deamination—but mechanistically is identical to *N*-dealkylation. The only difference between the two processes is that in *N*-dealkylation the nitrogen remains as part of the metabolized drug structure, while in deamination it does not. As mentioned earlier, if an aldehyde is formed it is subject to rapid oxidation to a carboxylic acid mediated by aldehyde dehydrogenase (Figure 7.25).

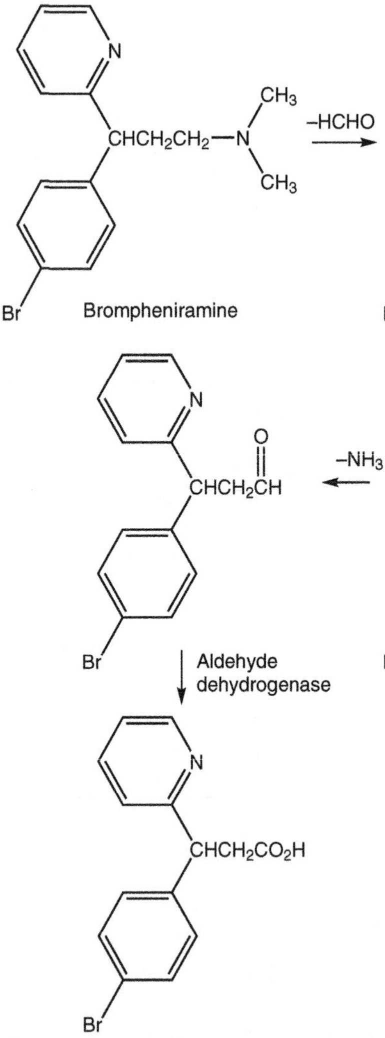

Figure 7.25 The antihistamine brompheniramine undergoes sequential *N*-dealklyation reactions to give first a secondary amine and then a primary amine. Since the carbon on the other side of the nitrogen also has two hydrogens it can undergo the same reaction, resulting in expulsion of the nitrogen as ammonia and giving rise to an aldehyde. This is finally oxidized by aldehyde dehydrogenase to a carboxylic acid.

In Chapter 6 it was mentioned that certain saturated nitrogen heterocycles such as piperidine and piperazine are common components of many drugs. These groups are cyclic amines and are also subject to metabolism by MFO by the mechanism outlined above. The amine and carbonyl components that result from this process, however, remain attached to each other by the remainder of the ring. This makes it likely that they will react with each other since amines are nucleophiles and aldehydes and ketones are electrophiles. Thus there are two possible outcomes:

• The amine can attack the carbonyl to form a carbinolamine, a process which is reversible. Carbinolamines, however, since they contain hydroxyl groups can be oxidized by alcohol dehydrogenase to form a new carbonyl group, but one that is attached to a nitrogen (a lactam).

• Alternatively, if an aldehyde was formed from MFO metabolism of the cyclic amine it may undergo direct oxidation catalyzed by aldehyde dehydrogenase to form a carboxylic acid. The acid can then condense with the amine to again form a lactam.

These processes are illustrated in Figure 7.26 for the attention-deficit hyperactivity disorder (ADHD) drug methylphenidate.

Figure 7.27 provides examples of O- and S-dealkylation. As mentioned above, this occurs by the exact same mechanism as N-dealkylation. Therefore, any alkyl group being removed by this process from oxygen or sulfur must also have at least one attached hydrogen atom.

Many drug molecules contain halogen substituents, usually with the halogen attached to an aromatic ring. As mentioned in the section on aromatic hydroxylation such substituents, being electron withdrawing, tend to slow the rate of hydroxylation relative to rings with electron-donating substituents. One class of drugs in which aliphatic halogen substitution is common is the inhalation anesthetics. Here the high degree of halogenation serves two main purposes—it decreases the flammability of these gaseous agents and also dramatically increases the log P so that they get absorbed rapidly through the lungs. Halothane is one such agent that is widely used (Figure 7.28). This simple, low-molecular-weight compound has only two carbons, one with three fluorines and the other with a chlorine, a bromine, and a hydrogen. This carbon is subject to oxidative metabolism by the MFO system. Oxygen gets inserted into the C–H bond to form an unstable intermediate. Electrons from the oxygen are pushed into the carbon and the better leaving group, which is bromide, is eliminated, resulting in a protonated acid chloride. Bromide can

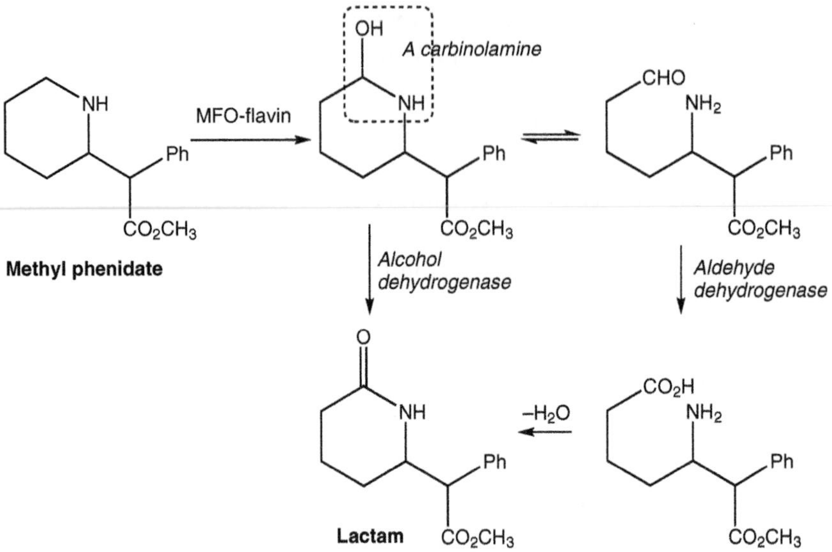

Figure 7.26 Methyl phenidate is used to treat ADHD. Its structure includes a cyclic amine. The process of metabolic dealkylation can occur on the unsubstituted position adjacent to the nitrogen giving rise to a carbinolamine, which can be hydrolyzed to an aldehyde and an amine. However, since the substrate was a cyclic amine these two new functional groups remain part of the same molecule. The aldehyde can be further oxidized to a carboxylic acid and this can react with the amine by elimination of water to form a lactam. Alternatively, the carbinolamine can be oxidized by alcohol dehydrogenase to directly give the same lactam, which is observed as a major metabolite of methyl phenidate.

remove the proton and the acid chloride reacts with water to eliminate HCl and form trifluoroacetic acid as a metabolite of halothane.

Chloroform is another compound that is metabolized by this pathway (Figure 7.29). This common organic solvent was used in the past as a general anesthetic until it was discovered to be hepatotoxic. Again oxygen inserts into the lone C–H bond and electrons from oxygen then displace one chloride ion with formation of the toxic gas phosgene, which is essentially a diacid chloride. Phosgene can react with water to form two molecules of HCl and carbonic acid, which decomposes into CO_2 and water. Alternatively, cellular macromolecules with nucleophilic groups can react at the carbonyl and displace chloride to form a new covalent bond between the carbonyl and the macromolecule. This can happen a second time with another nucleophile giving rise to *cross-linking*, which is very difficult for the body to repair.

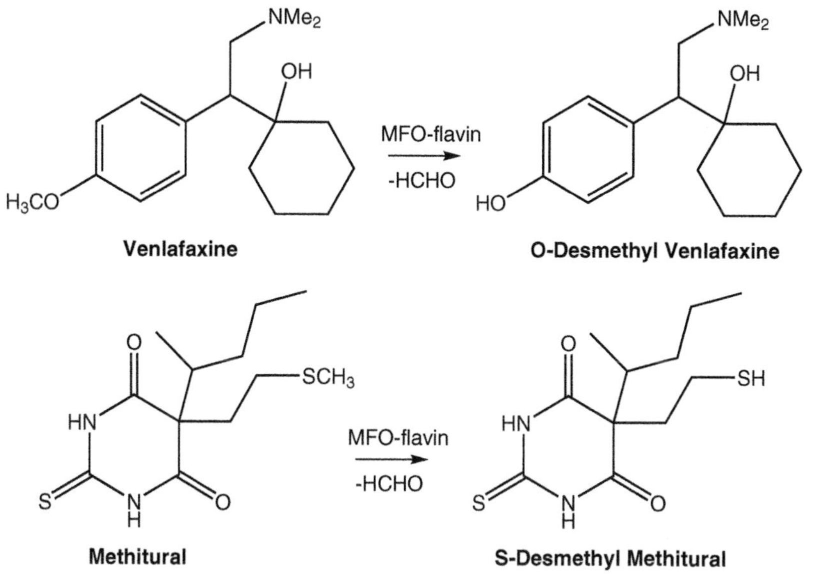

Venlafaxine

O-Desmethyl Venlafaxine

Methitural

S-Desmethyl Methitural

Figure 7.27 Dealkylation on *O* and *S* is also a common metabolic pathway which operates by the same mechanism as *N*-dealkylation. *Top*: an example of *O*-demethylation is seen with the antidepressant venlafaxine. *Bottom*: the anesthetic methitural provides an example of *S*-demethylation.

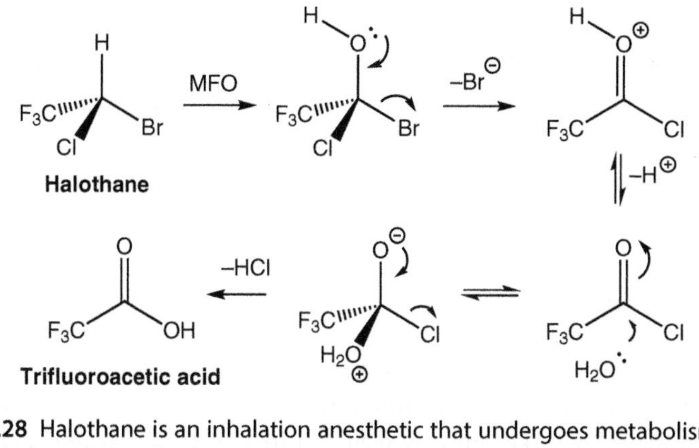

Halothane

Trifluoroacetic acid

Figure 7.28 Halothane is an inhalation anesthetic that undergoes metabolism by oxidative dehalogenation. Oxygen inserts into the C–H bond and bromide is eliminated to form an acid chloride. This reacts with water to form trifluoroacetic acid.

One final oxidative metabolism that can occur is desulfurization. Oxidative desulfurization, while not well understood mechanistically, involves the conversion of a thiocarbonyl group (C=S) into a regular carbonyl (C=O). A well-known example is the metabolic conversion of the

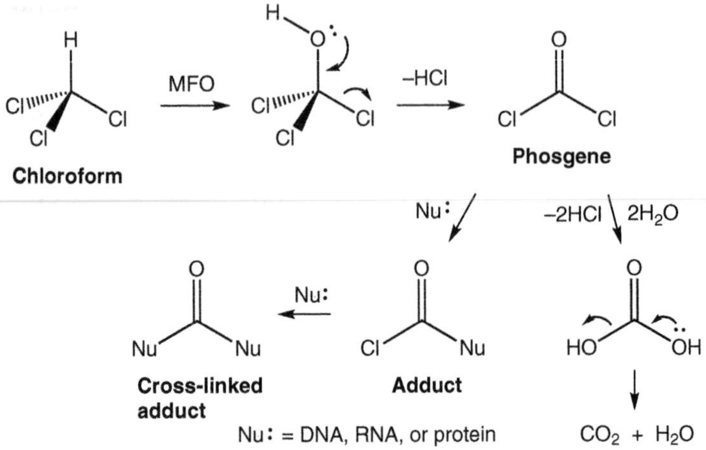

Figure 7.29 Chloroform is also metabolized by oxidative dehalogenation to form the toxic compound phosgene. While this can react with water to form carbonic acid that breaks down into carbon dioxide and water it can also react with nucleophilic groups in proteins and nucleic acids. If it reacts twice, the result is a cross-linked adduct which is very difficult for the body to repair.

anesthetic thiopental into pentobarbital (Figure 7.30). Thiopental is the conjugate acid of sodium pentothal (truth serum). Although the actual mechanism of the metabolic transformation is not known with certainty, a plausible route would involve oxidation of the sulfur by the flavin mono-oxygenase system to give a zwitterionic intermediate. Attack by water would then result in a species having carbon bonded to an SR and OH group, a monothioacetal. Such structures are readily hydrolyzed to give carbonyl derivatives.

PHASE I: REDUCTIONS

A second category of phase I metabolic transformations that can occur is reduction. Most commonly, metabolic reductions are associated with functional groups such as ketones, as well as nitro and azo groups. Less commonly aldehydes and sulfoxides can also undergo metabolic reduction. *Reduction* is the process of lowering the oxidation state of a group by the addition of electrons. Most commonly these electrons are added as a hydride ion, H^-. Usually reduction is also accompanied by a change in hybridization from sp to sp^2 or from sp^2 to sp^3. Thus a $C=O$ bond is reduced to CH–OH.

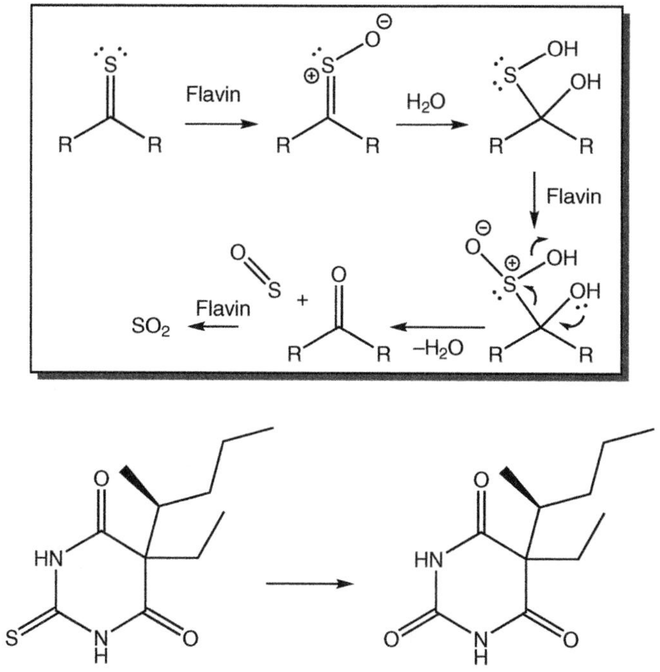

Figure 7.30 *Top*: a plausible mechanism for the oxidative conversion of thiocarbonyl groups into regular carbonyls. *Bottom*: the thiobarbiturate thiopental is converted metabolically into pentobarbital.

In Chapter 3 it was shown that there are a large number of functional groups that contain a carbonyl bond. Of these, however, only ketones and to a lesser extent aldehydes typically undergo metabolic reduction. Aldehydes are more commonly oxidized further to carboxylic acids as described above, and their metabolic reduction to primary alcohols is relatively rare (Figure 7.31). Ketones, however, being inert to further oxidation undergo ready metabolic reduction to secondary alcohols.

The enzymes that are responsible for reducing ketones and aldehydes are known collectively as the aldo–keto reductases. These enzymes require NADPH as a cofactor. The mechanism for these transformations is shown in Figure 7.32. NADPH is a 1,4-dihydropyridinium derivative that has a positively charged nitrogen and is not aromatic. In the presence of a reducible group such as a ketone, a base removes the proton from the ring nitrogen of NADPH and both electrons from the N–H bond are pushed into the heterocyclic ring. These force a pair of electrons from one of the

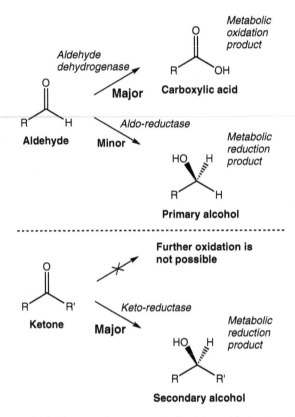

Figure 7.31 *Top*: aldehydes usually are metabolized by oxidation to carboxylic acids. Occasionally reduction mediated by aldo-reductase can give rise to primary alcohols. *Bottom*: ketones are resistant to further oxidation. Their major route of metabolism is reduction by keto-reductase to give secondary alcohols.

existing π bonds toward the 4-position which contains two hydrogen atoms. As there can be only four bonds to carbon, something has to give and it is one of the C–H bonds that break, with both electrons going to the leaving hydrogen to form a hydride anion. The ring is now aromatic (a pyridinium ring) and is called NADP$^+$. The released hydride anion, which is a nucleophile, attacks the carbonyl carbon of the ketone, which pushes electrons from the C=O π bond onto the oxygen to form an alk-oxide anion. This reacts with any convenient proton source, such as the one released from NADPH originally, to complete the formation of a secondary alcohol.

Examples of the metabolic reduction of naloxone, a ketone, and chloral hydrate, which forms an aldehyde, are shown in Figure 7.33.

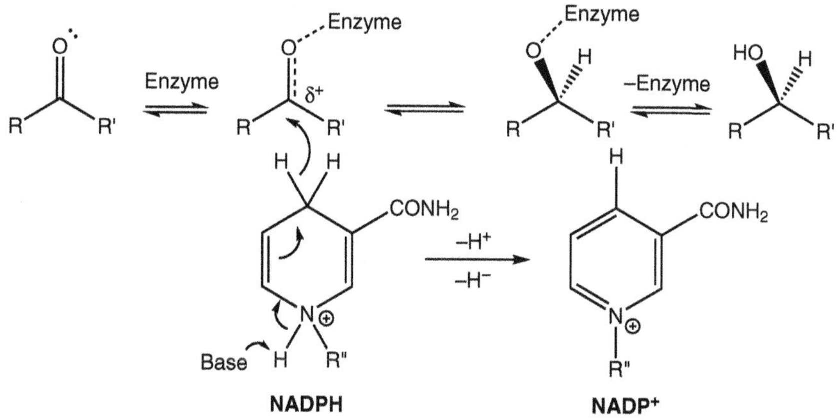

Figure 7.32 The mechanism of reduction of ketones and aldehydes utilizes NADPH. Coordination of the enzyme to the carbonyl increases the amount of positive charge on carbon. When a base removes a proton from NADPH the electrons from the N–H bond are forced into the ring leading to the expulsion of one of the hydrogens at the 4-position in the form of a nucleophilic hydride anion. This adds to the carbonyl to achieve the reduction and give the corresponding alcohol.

Nitro groups are another class of functionality that readily undergo metabolic reduction. The reduction of a nitro group usually results in formation of a primary amine, although other products are sometimes produced. These reductions take place in a stepwise manner and are catalyzed by a flavin-dependent NADPH-cytochrome P450 reductase. The mechanism is shown in Figures 7.34 and 7.35. The first of these figures shows how the flavin moiety becomes activated so that it can serve as a source for single electrons [5]. The oxidized form of flavin is reduced by stepwise addition of two electrons to form a reduced form of flavin. In the presence of a suitable acceptor molecule, electrons flow from the top nitrogen to form a new imine bond. In the process one electron is donated to the acceptor to form a radical anion. A base removes hydrogen from the central nitrogen and the electrons from the N–H bond then force donation of the remaining unpaired electron to reform oxidized flavin and pair up the electron on the acceptor to form an anion.

It has been shown that reduction of nitro groups proceeds through a free-radical mechanism in a stepwise manner (Figure 7.35) [6]. An electron from flavin attacks the double-bonded oxygen, displacing a pair of electrons onto the nitrogen. Donation of a second electron then follows to complete a pair of electrons on this oxygen. Protonation of both

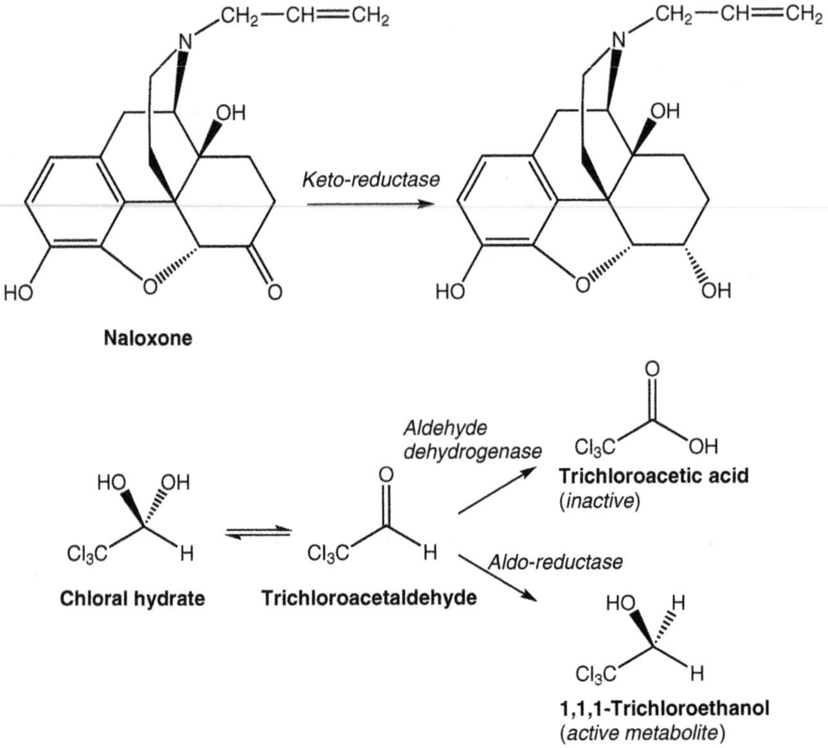

Figure 7.33 *Top*: an example of metabolic reduction of a ketone is seen for the narcotic antagonist naloxone. *Bottom*: chloral hydrate (the active ingredient in a "Mickey Finn") exists in equilibrium with trichloroacetaldehyde. Some of this is oxidized to trichloroacetic acid which is not active as a hypnotic agent. The remainder gets metabolically reduced to 1,1,1-trichloroethanol, which is the active sleep-inducing agent.

oxygens and elimination of water then gives a *nitroso* compound as an intermediate. The process then repeats with two more electrons added to the oxygen, displacing a pair of electrons onto nitrogen, which then becomes protonated to form an N–H bond. Protonation of the oxygen allows for water to be eliminated, leaving behind an electron-deficient nitrogen known as a *nitrene*. The addition of two more electrons and protonation of the nitrogen gives the primary amine.

Thus metabolic nitro reduction proceeds through a nitroso compound to a hydroxylamine, and finally to a primary amine. Occasionally, the nitroso compound and/or the hydroxylamine can be detected as a metabolite, but most commonly it is the amine that is the major metabolite of nitro

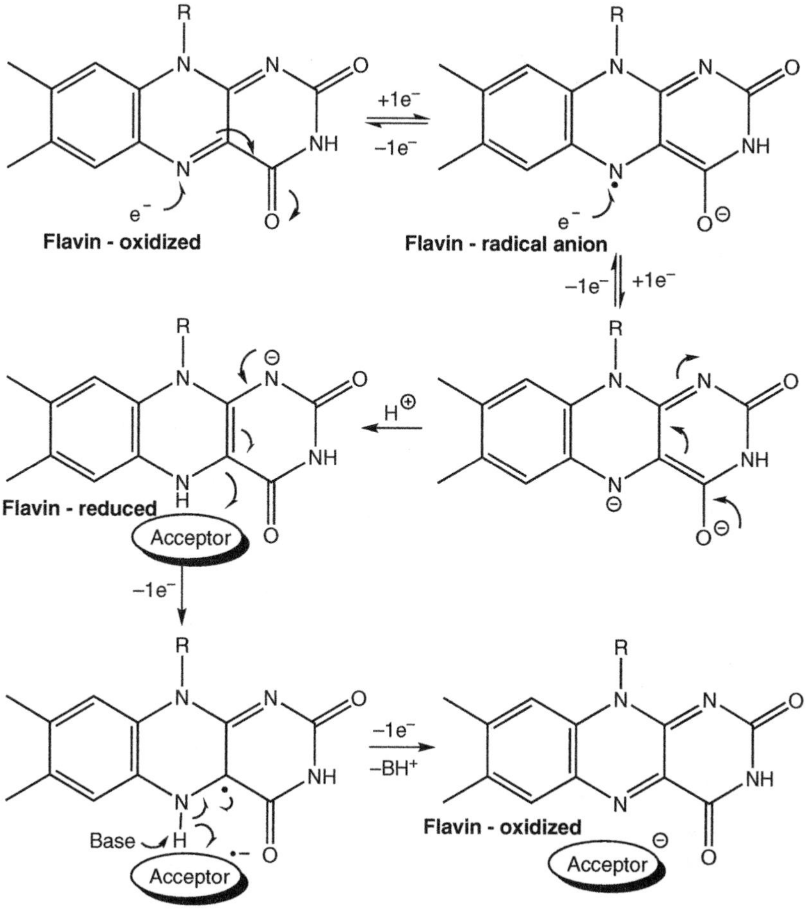

Figure 7.34 The mechanism of metabolic nitro reduction involves the donation of unpaired electrons to the nitro group. Here the production of unpaired electrons from flavin is shown. The acceptor in this diagram would be a nitro group [5].

reduction. An example of a metabolic nitro reduction is shown in Figure 7.36 for the muscle relaxant, dantrolene.

Azo compounds, while not as prevalent in drugs as nitro groups, are an essential part of an important group of nonsteroidal antiinflammatory drugs (NSAIDS). An azo compound is one that contains an N=N bond, and these groups are subject to metabolic reduction. Many of the azo NSAIDS are poorly absorbed when taken orally and reach the intestines largely unaltered. The human intestine is alive with bacteria and these are rich sources

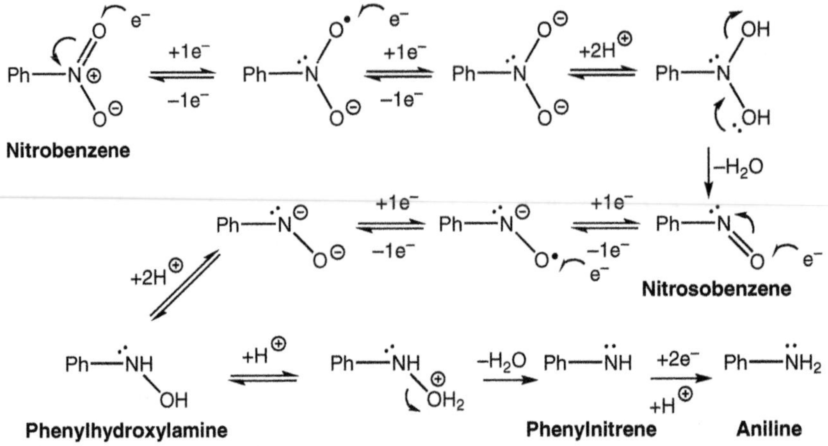

Figure 7.35 The stepwise reduction of a nitro group is demonstrated for nitrobenzene. Note that addition of a lone electron to the oxygen displaces a pair of electrons onto the nitrogen. After two electrons and two protons are added and a molecule of water is eliminated a nitroso compound is the first intermediate. These are sometimes isolated among the metabolites of nitro compounds. Further addition of two more electrons and two protons gives a hydroxylamine. Finally, elimination of water leaves an electron-deficient nitrogen species known as a nitrene. Two more electrons are added to give an amide anion that is protonated to give the final primary amine [6].

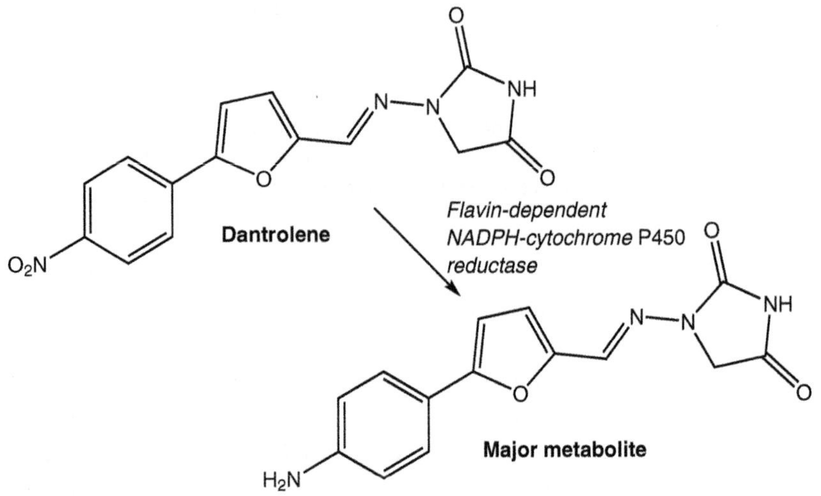

Figure 7.36 The muscle relaxant dantrolene is metabolized by reduction of the nitro group to an aryl amine.

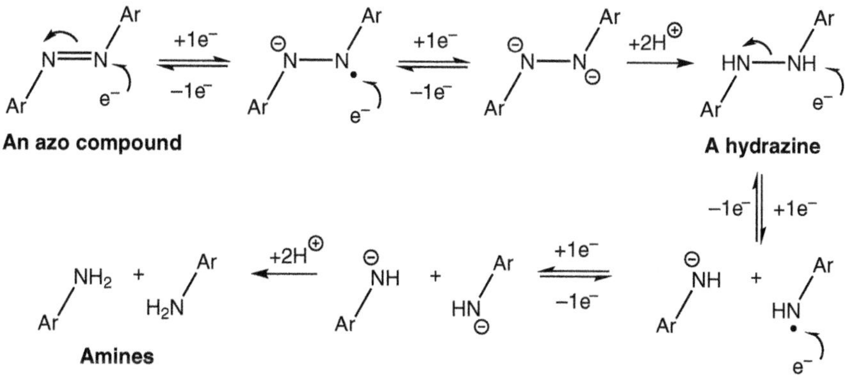

An azo compound

A hydrazine

Amines

Figure 7.37 Azo compounds are also metabolized via addition of unpaired electrons. Addition of the first electron breaks the π bond between the two nitrogens to form a radical anion. A second electron and two protons are then added to produce a hydrazine. With the addition of another electron the N–N bond breaks. Again, one more electron and two protons are added to give the final products, which are two amines [7].

of nitro reductase enzymes which are capable of reducing an azo compound into two amines. As was the case for nitro reduction, the reduction of azo compounds is also a stepwise process. The mechanism is depicted in Figure 7.37. Again, a radical mechanism is believed to operate with a single electron being added to one of the nitrogens, pushing an electron pair to the other nitrogen that then becomes protonated [7]. One more electron adds to the N-based radical and this too gets protonated. The new functional group that forms as an intermediate has two sp^3 nitrogens connected by a σ bond and is called a *hydrazine*, or *hydrazo* compound. In the second part of the mechanism, another electron is added to one of the nitrogens with simultaneous cleavage of the N–N bond. The electrons from that bond move onto the second nitrogen and it becomes protonated to form an amine. The nitrogen radical then acquires one more electron to become anionic. Protonation gives a second molecule of amine. An example of this is shown in Figure 7.38 for the symmetrical compound olsalazine, an NSAID used to treat chronic inflammatory bowel disease. Here the parent olsalazine is considered to be a prodrug. After metabolic reduction two molecules of 5-aminosalicylic acid are produced. This compound is the actual agent that alleviates the inflammatory condition.

Sulfoxides, like aldehydes, are usually more likely to undergo metabolic oxidation. Occasionally, however, sulfoxides can be metabolically reduced to sulfides. A plausible mechanism for such a process is shown at the top of Figure 7.39. Protonation of the (partially) negatively charged sulfoxide

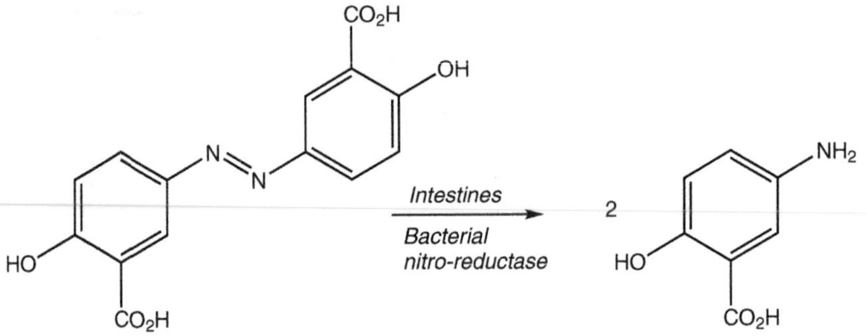

Olsalazine
(prodrug)

5-Aminosalicyclic acid
(active metabolite)

Figure 7.38 An entire subclass of NSAID prodrugs has been designed for treating chronic inflammatory bowel disease. These possess a central azo group and are not substantially metabolized until they reach the intestines. There bacteria, which are rich in nitroreductase enzymes, metabolize the azo bond, breaking it to form 5-aminosalicylic acid which is the active drug that combats the inflammation. An example is the symmetrical compound olsalazine which releases two molecules of the active amine on metabolic reduction.

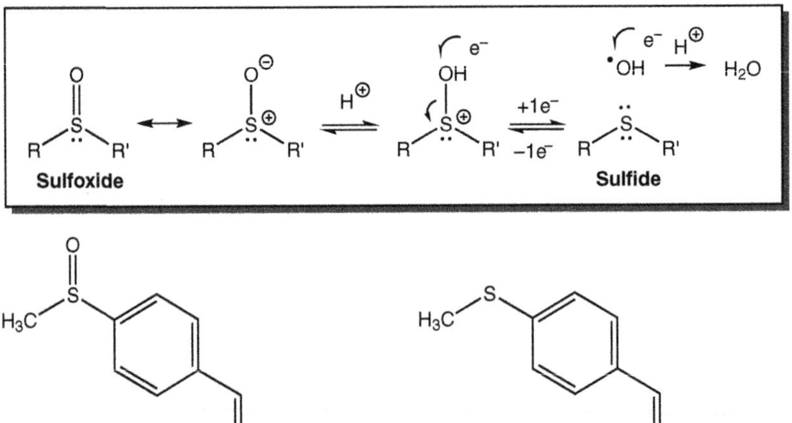

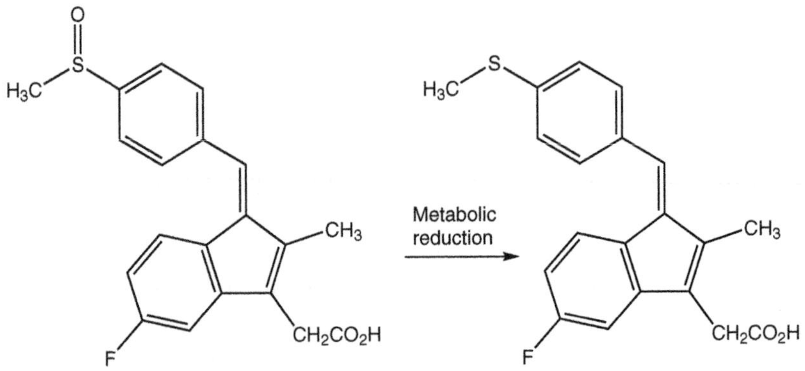

Sulindac
(prodrug)

Sulfide
(active metabolite)

Figure 7.39 Sulfoxides, like aldehydes, are more likely to undergo metabolic oxidation than reduction. A plausible mechanism for the metabolic reduction of sulfoxides to sulfides is presented in the top of this figure. Sulindac is a rare example of a drug that undergoes metabolic reduction of a sulfoxide to a sulfide. In this case sulindac is a prodrug because it is inactive in its native form. The sulfide metabolite is the active agent that is effective against inflammation.

oxygen gives rise to a positively charged *sulfonium ion*. Addition of an electron to the oxygen would break the S–O bond to form the sulfide and hydroxyl radical. This can add one more electron to form hydroxide anion which would be protonated to form water. An example of metabolic sulfoxide reduction occurs with the NSAID Sulindac (Figure 7.39). Sulindac is a prodrug with a sulfoxide group. The sulfoxide, which is inactive, undergoes both metabolic oxidation to form a sulfone, which is also inactive, and reduction to form the methyl sulfide, which is active as an antiinflammatory agent.

PHASE I: HYDROLYSIS

The third major class of phase I metabolic transformations is *hydrolysis*, which formally is the process of breaking bonds by the addition of water. Those functional groups that are most often metabolized via hydrolysis include esters (and lactones) and amides (and lactams). Hydrolysis of ordinary esters always produces two molecules, a carboxylic acid and an alcohol (Figure 7.40). For lactones, because they are cyclic esters, upon hydrolysis

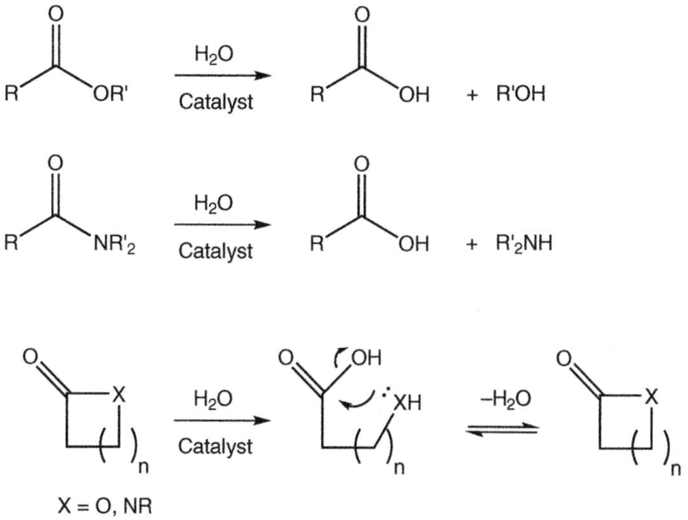

Figure 7.40 Hydrolysis of esters produces carboxylic acids and alcohols, whereas amides are hydrolyzed to carboxylic acids and amines. Lactones and lactams are hydrolyzed to give compounds having a carboxylic acid group at one end and an alcohol (or amine) at the other. Occasionally, these groups can recombine to reform the lactone or lactam, with no hint that hydrolysis had occurred.

the carboxylic acid group and the alcohol remain attached as part of the same molecule. Amides also undergo hydrolysis to form two molecules, one of which is again a carboxylic acid, while the other is a primary or secondary amine. As with lactones, hydrolysis of lactams produces only a single molecule that contains a carboxylic acid group and an amine that are joined by a chain of atoms. The ring-opened hydrolysis products of lactones and lactams may recyclize with elimination of water if conditions are favorable.

The mechanism of a hydrolysis reaction requires attack by water (a nucleophile) at the carbonyl carbon, which is an electrophilic site. Remember that in a carbonyl group the carbon carries a partial positive charge (see Chapter 3). With functional groups such as aldehydes and ketones the charge on that carbon is not counter-balanced by any sources of electron density and is subsequently highly electrophilic. This means that nucleophiles, including water, will readily attack at this position. Reaction of water with an aldehyde for example (Figure 7.41) creates a new C–OH bond and transfers electron density from the π bond onto the existing oxygen, which then becomes protonated. The result is a functional group known as a *hydrate*, which has two OH groups attached to a single carbon. Hydrates, however, exist in equilibrium with their original aldehyde forms via elimination of a water molecule. That is because breaking a C–OH bond is much easier than breaking an attached C–C or C–H bond, especially when the OH group is protonated so that it can be eliminated as water.

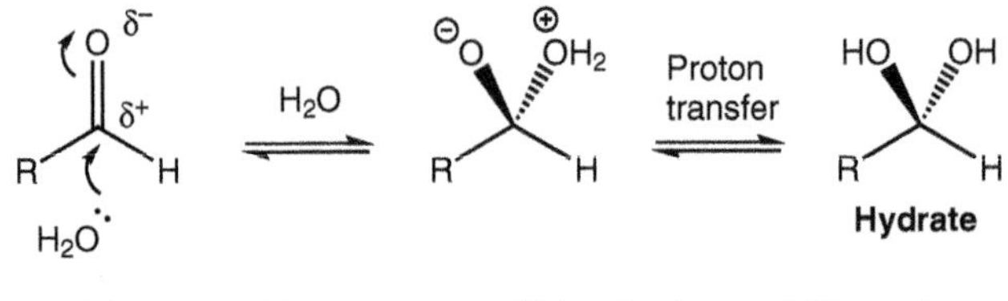

Figure 7.41 Aldehydes and ketones are sufficiently electrophilic so that nucleophiles such as water can attack the carbonyl to give rise to hydrates. While the reactions are reversible, it is only the OH group that can be eliminated because it is weaker than the C–C or C–H bonds.

With esters and amides, however, the heteroatom attached to the carbonyl carbon can donate electrons which has the effect of decreasing the charge on the carbon and reducing its electrophilicity (Figure 7.42). In order for water to successfully attack the carbon the amount of charge

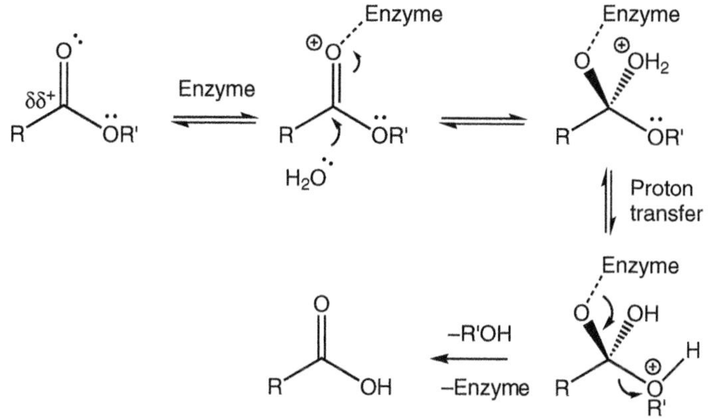

Figure 7.42 Esters and amides have a heteroatom with lone pair electrons attached to the carbonyl group. This diminishes the amount of positive charge on the carbon making it less electrophilic. For a nucleophile such as water to attack a catalyst is required to bind to the oxygen, resulting in an increase in positive charge on the carbon. Metabolically this is achieved with a hydrolase enzyme acting as the catalyst. Again the attack of water is reversible. A proton, however, can be transferred to the OR (or NR$_2$) group, which converts it into a good leaving group. When the enzyme dissociates from the oxygen the electrons help push out the leaving group to form a carboxylic acid and an alcohol (or amine).

needs to be increased. This is achieved by having an enzyme coordinate with one of the lone pairs on the carbonyl oxygen. The resonance structure with a formal positive charge on carbon is more stable than that bearing a positive charged on oxygen. Now water can attack to form a tetrahedral intermediate, with carbon attached to an R group, alkoxy group, OH_2^+ (the attacking water), and an O-enzyme. Formation of this intermediate is reversible. However, transfer of a proton from OH_2^+ to the OR group can also occur. Now the tetrahedral intermediate has an OH and a HOR^+ in addition to the other two groups. If the enzyme dissociates away from the oxygen, those electrons will be released back toward carbon and the protonated alkoxy group will be forced out—that is, the HOR^+ becomes what is known as a *leaving group*. Thus an alcohol is produced along with a carboxylic acid. The mechanism for amide hydrolysis is identical except that the nitrogen is a much better electron-donating group which decreases the electrophilicity of the carbon to a greater extent than did the ester alkoxy group. This is because nitrogen can better tolerate having a positive charge than the more electronegative oxygen. Thus the hydrolysis of amides tends to be much slower than that of esters. Both processes, however, are common metabolic pathways. The enzymes that catalyze these processes are called, respectively, esterases and amidases, and are prevalent in plasma and other body tissues. Figure 7.43 shows two examples of metabolic hydrolysis.

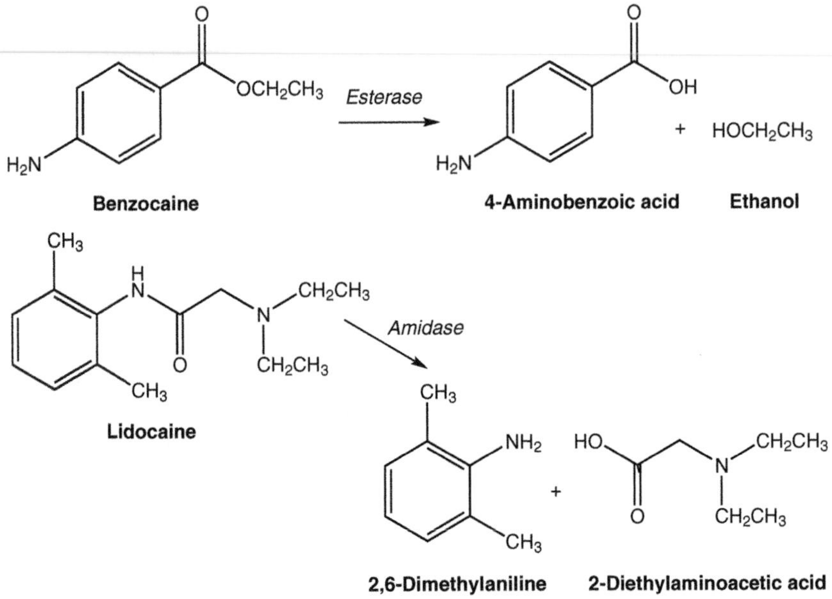

Figure 7.43 *Top*: benzocaine, a local anesthetic, is an example of an ester-containing drug that is metabolized by hydrolysis. *Bottom*: lidocaine, another anesthetic, contains an amide group and is metabolized by amidase to 2,6-dimethylaniline and 2-diethyl-aminoacetic acid.

Ester and amide hydrolysis are strongly influenced by steric factors. From the mechanism of hydrolysis shown above it is seen that while the parent carbonyl derivatives are sp^2 hybridized at the carbon, hydrolysis proceeds through a tetrahedral intermediate. Thus bond angles go from a nominal 120° in the ester or amide to about 109° in the intermediate. Crowding therefore increases before one group gets eliminated and the hybridization returns to sp^2. Large substituents that are close to the carbonyl group can influence the rate at which metabolic hydrolysis occurs. This is often used to advantage in drug design to modify the duration of action of ester- and amide-containing drugs. If a long duration is required then bulky groups are used on the alkoxy portion or adjacent to the carbonyl group, whereas if only a short duration is desired, then small groups and easy accessibility to the ester groups ensure that rapid hydrolysis will occur (Figure 7.44).

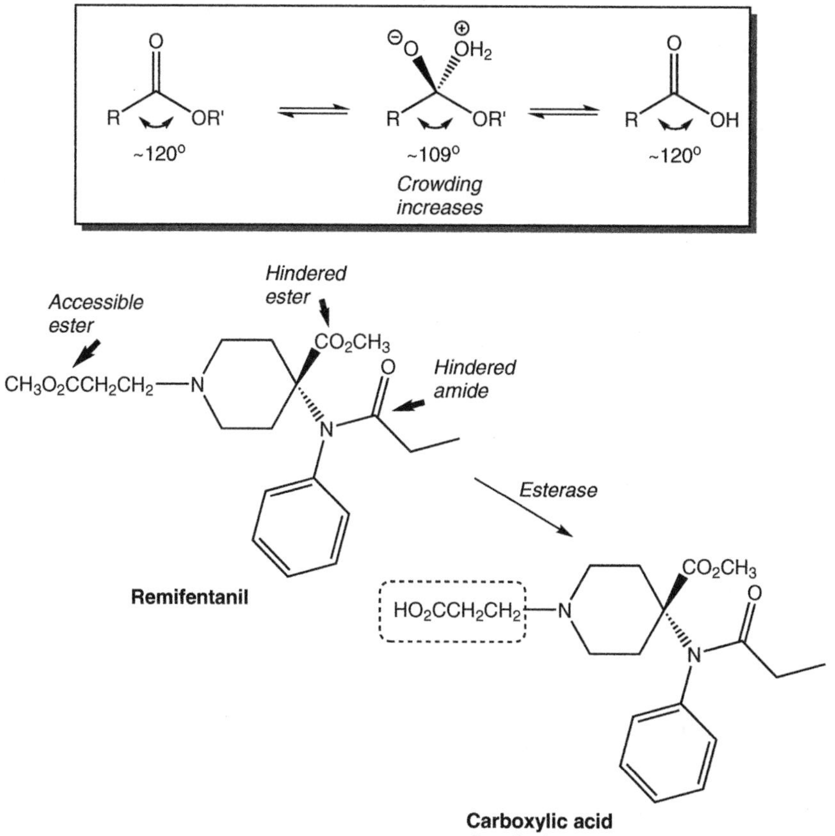

Figure 7.44 *Top*: demonstration of how steric crowding increases in the tetrahedral intermediate formed during ester hydrolysis. *Bottom*: the ultra-short-acting anesthetic remifentanil is a good example of using structure to design drugs with certain properties. Here an unhindered methyl ester undergoes rapid hydrolysis by esterase to form an inactive carboxylic acid giving the drug a very short half-life. Note that neither the hindered ester nor the amide are metabolized as rapidly.

PHASE II: CONJUGATION

All the metabolic transformations discussed to this point belong to phase I, the purpose of which is to alter the chemistry of a xenobiotic molecule so as to terminate its normal activity, usually rendering the molecule more polar than when it entered the body. However, simply adding a single OH or NH_2 group to a large molecule is generally not a sufficient structural change to alter its log P enough so that the molecule becomes water soluble and can be excreted. This is the job of phase II metabolism. In phase II, parent compounds and phase I metabolites are generally altered by attaching a highly polar group to appropriate functionality. The resulting modified compounds thereby gain sufficient polarity to render them water soluble so they can be excreted, usually in the urine. The process of attaching another molecule to a substrate is known as conjugation. Among the more common classes of phase II metabolism are:

- Glucuronide conjugation
- Sulfate conjugation
- Glycine and glutamate conjugation
- Glutathione (GSH) conjugation
- Acetylation
- Methylation

In this section of the chapter we will explore conjugation reactions beginning with glucuronide formation. Glucuronides are among the most common of phase II metabolites. Functional groups that are susceptible to conjugation with glucuronic acid include alcohols, phenols, carboxylic acids, amines, and thiols.

β-**Glucuronic acid**

Glucuronic acid is made in the body from the highly abundant sugar D-glucose (Figure 7.45). Glucose is often converted in the body into a phosphate ester, α-D-glucose-1-phosphate. This is a substrate for the enzyme uridine triphosphate phosphorylase, which appends a uridine-5′-phosphate to the existing phosphate to form uridine-5′-diphosphate-α-D-glucose (UDPG). This in turn is a substrate for another enzyme, UDPG dehydrogenase, which oxidizes the primary alcohol to the carboxylic acid, uridine-5′-diphospho-α-D-glucuronic acid. Examination of this structure reveals a tetrahydropyran (cyclic six-membered ether) having a carbon that is attached to the ring oxygen also attached to another OR group, in this case an O-diphosphate-uridine. This creates a functional group known as an *acetal*. Acetals (sp^3 carbons bearing two OR groups) are derived from aldehydes and are subject to attack by nucleophiles in

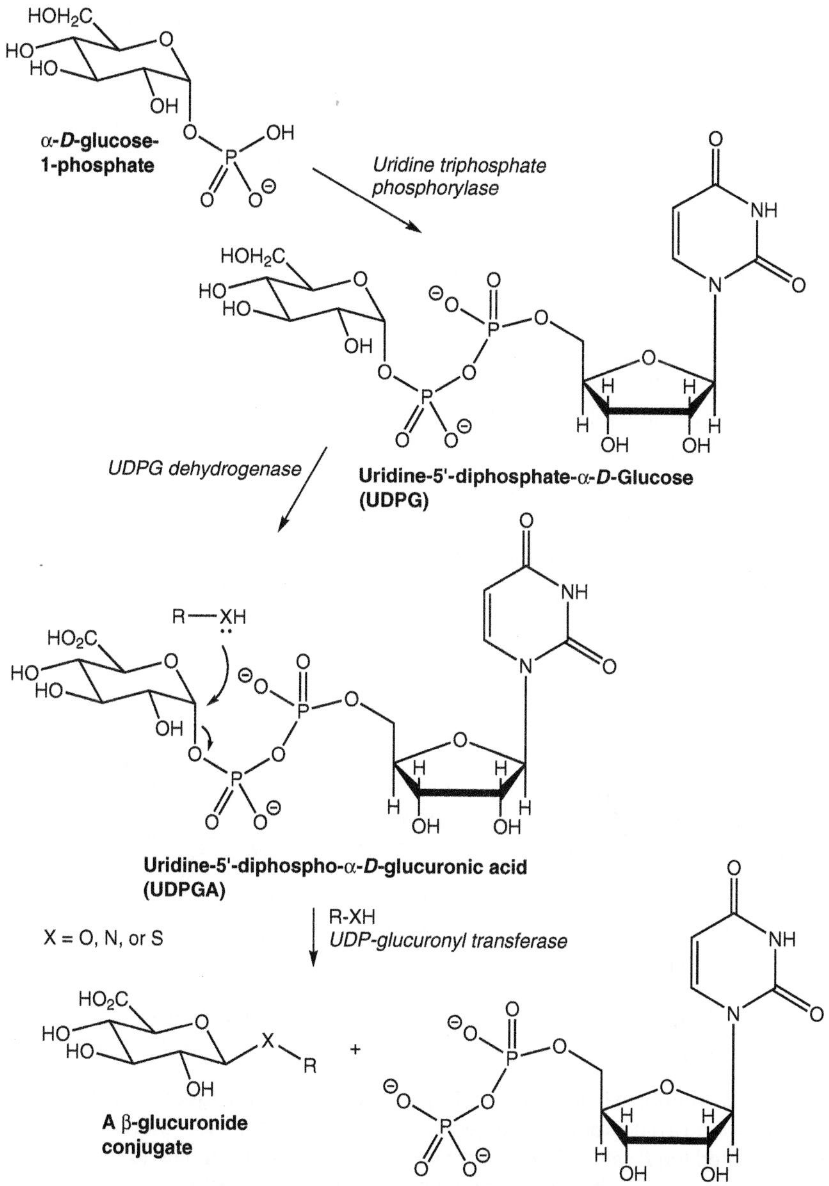

Figure 7.45 The formation of conjugates of glucuronic acid begins with α-D-glucose. Glucose is often converted in the body into its 1-phosphate. When needed, the phosphate becomes phosphorylated by a uridine phosphate to form UDPG. The primary alcohol on the glucose ring is then oxidized by UDPG dehydrogenase to form a phosphorylated glucuronic acid. Nucleophilic functional groups such as alcohols, phenols, carboxylic acids, amines, and thiols react at the acetal carbon of the sugar, inverting the configuration at that site to give a β-glucuronic acid conjugate (glucuronide). This is catalyzed by UDP-glucuronyl transferase.

the presence of a catalyst, with displacement of one of the OR groups. In this case the diphosphate-uridine group renders the oxygen to which it is attached into a good leaving group. Thus in the presence of a nucleophile such as an alcohol, amine, carboxylic acid, or thiol, the enzyme uridine-5'-diphosphate-glucuronyl transferase catalyzes the displacement of the O-diphosphate-uridine group to form a β-glucuronide conjugate.

Morphine is an example of a parent compound that undergoes extensive glucuronide conjugation on the phenol (Figure 7.46). Note the change in the calculated log P values for morphine (Clog$P$ = +0.57) and its 3-glucuronide conjugate (Clog$P$ = −3.68). Thus attaching the highly polar sugar group with its three secondary alcohols and one carboxylic acid group makes the compound highly water soluble allowing for its excretion in the urine. Figure 7.47 shows the sequential phase I and, phase II metabolism of diazepam. The parent compound undergoes α-oxidation at the 3-position to form an alcohol. The OH group lowers the Clog$P$ value from 2.96 for the parent compound to 2.34. In phase II, the alcohol is converted into a glucuronide conjugate and it is seen that the Clog$P$ value decreases substantially to 0.76.

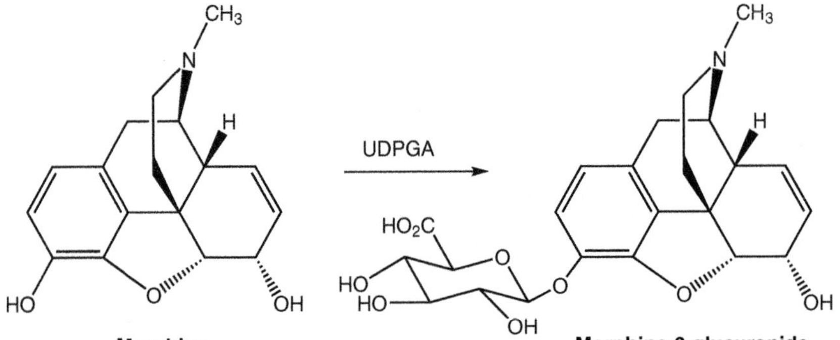

Morphine
Clog$P$: 0.57

Morphine-3-glucuronide
Clog$P$: −3.68

Figure 7.46 Morphine is an example of a parent drug that undergoes phase II metabolism by forming a glucuronide conjugate from its phenol group. Note that in the process the calculated log P value (Clog$P$) decreases from +0.57 for morphine to −3.68 for the conjugate, making it substantially more water soluble. (Clog$P$ values were determined using ChemBioDraw Ultra 13.0, Perkin Elmer Informatics, Waltham, MA).

Sulfate conjugates are formed mainly from phenols, although they can also be formed from alcohols and aromatic amines. In the body sulfate is less available than glucuronic acid and so these conjugates are generally formed to a lesser extent than glucuronides. Sulfate needs to

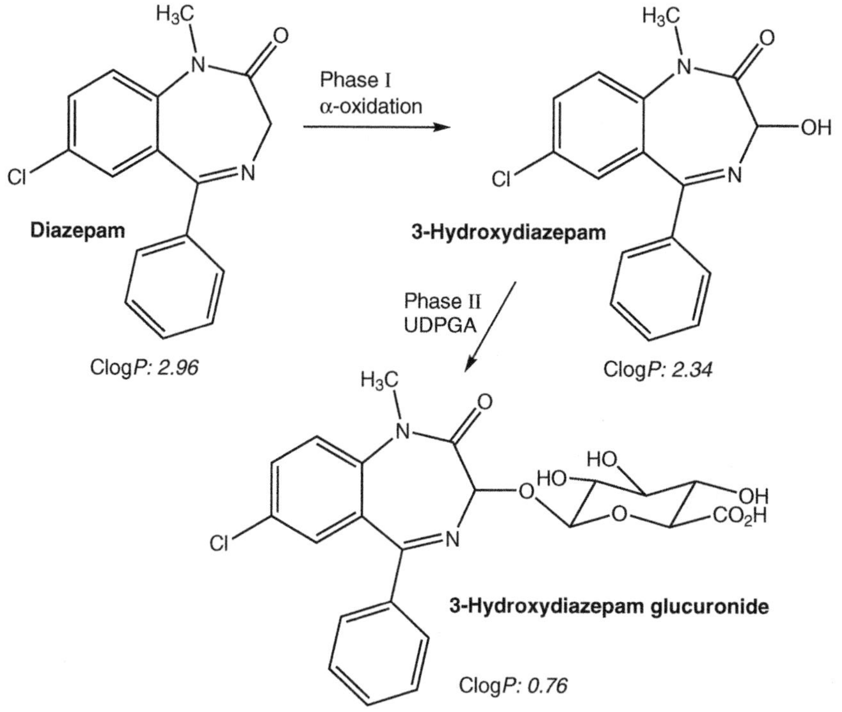

Figure 7.47 Diazepam is an example of a drug that first undergoes phase I α-oxidation to its 3-hydroxy metabolite and then phase II glucuronide conjugation of the alcohol. Addition of the alcohol group in phase I lowers the ClogP value only slightly from +2.96 to +2.34. Conjugation with glucuronic acid, however, lowers the ClogP substantially to +0.76. (ClogP values were determined using ChemBioDraw Ultra 13.0, Perkin Elmer Informatics, Waltham, MA).

be activated before it can react with nucleophiles. This is done by attaching a good leaving group to the sulfate moiety. This is an adenosine phosphate group and the resulting cofactor is called 3′-phosphoadenosine-5′-phosphosulfate (Figure 7.48). The enzyme sulfotransferase catalyzes the attack of nucleophilic groups on the sulfur, releasing 3′,5′-adenosine diphosphate and producing the sulfate conjugate of the nucleophile. Acetaminophen is an example of a drug that forms a sulfate conjugate, particularly in children in whom this is the major urinary metabolite. This is shown in Figure 7.49. As with glucuronides, the sulfate conjugate has a substantially lower ClogP value (−0.84) as compared to the parent compound (ClogP = +0.49).

The amino acids glycine and glutamine form conjugates with carboxylic acids. The products are known as carboxyamides and are more water soluble than the original carboxylic acids. Chemically the reaction is attack of a

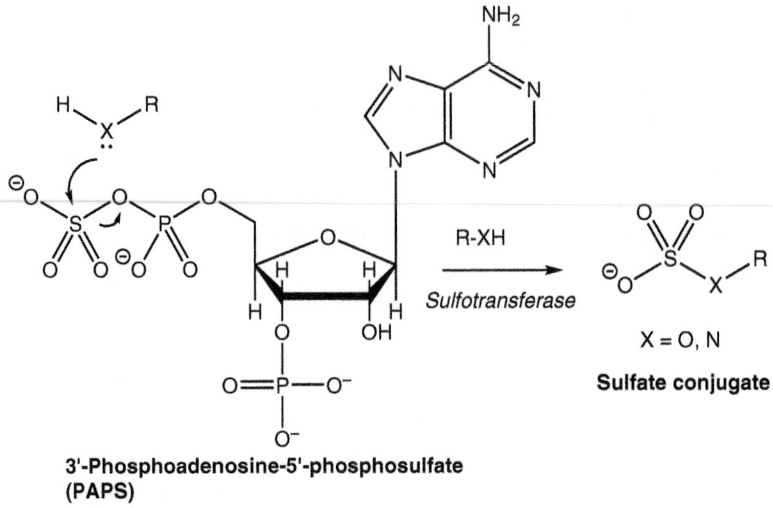

3'-Phosphoadenosine-5'-phosphosulfate (PAPS)

Figure 7.48 Phenols and occasionally alcohols and aromatic amines can serve as nucleophiles in reactions with 3'-phosphoadenosine-5'-phosphosulfate (PAPS). They attack the sulfur displacing an adenosine diphosphate to form, after loss of a proton, a sulfate conjugate. The process is catalyzed by sulfotransferase.

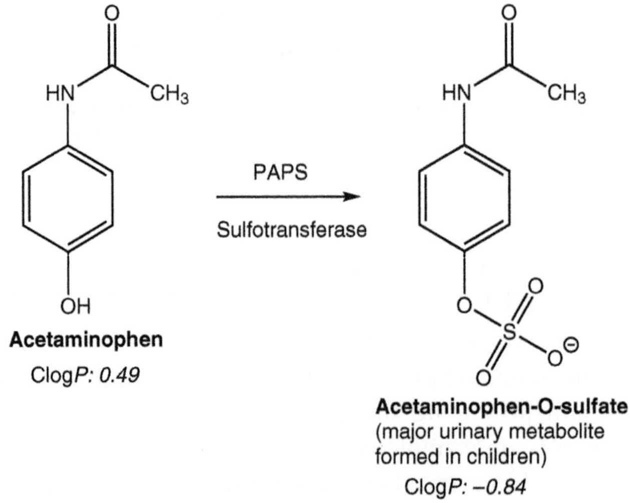

Acetaminophen
ClogP: 0.49

Acetaminophen-O-sulfate
(major urinary metabolite formed in children)
ClogP: −0.84

Figure 7.49 Acetaminophen has a phenol group that is subject to phase II metabolic conversion into a sulfate conjugate. This changes the ClogP value from +0.49 in the drug to −0.84, increasing the water solubility. This is the major urinary metabolite found in children in whom acetaminophen was administered. (ClogP values were determined using ChemBioDraw Ultra 13.0, Perkin Elmer Informatics, Waltham, MA).

nucleophile (the amino group from the amino acid) on the carbonyl of the carboxylic acid with displacement of OH. For this to be favorable the OH must be activated, since OH$^-$ is a relatively poor leaving group. This is done in two steps (Figure 7.50). First the acid is phosphorylated by adenosine triphosphate to form an acyl monophosphate. This then reacts with the

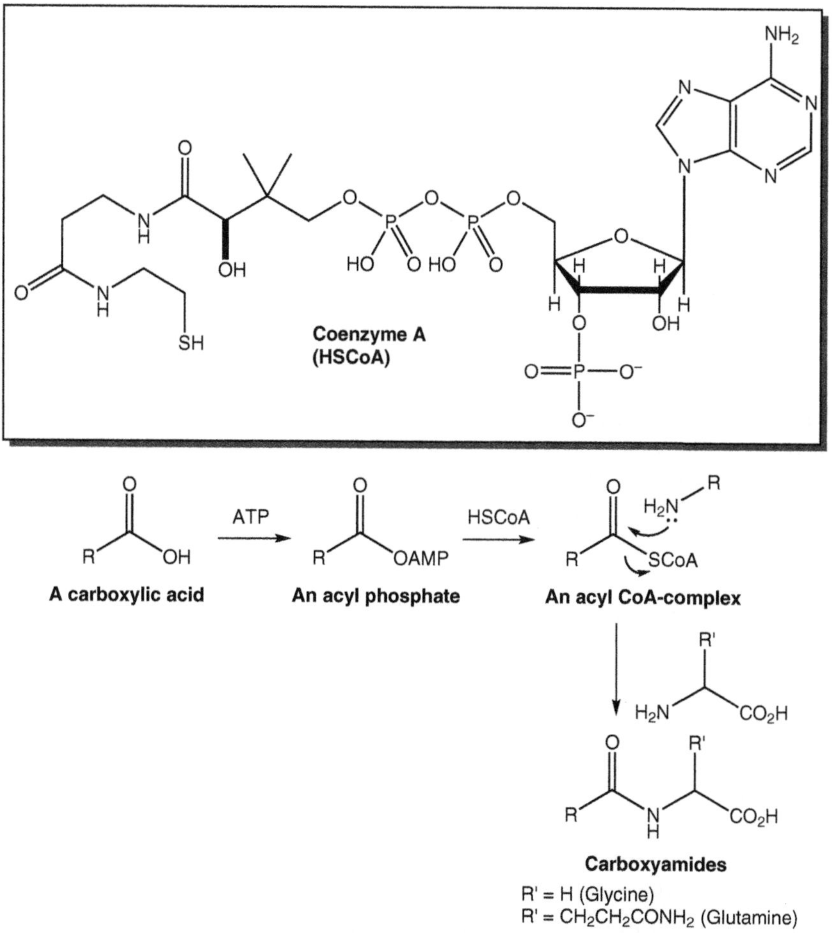

Figure 7.50 *Top*: structure of coenzyme A, showing the terminal thiol group, which is its most nucleophilic site. *Bottom*: carboxylic acids are converted by adenosine triphosphate (ATP) into acyl monophosphates. Coenzyme A displaces the phosphate group to form an acyl-CoA complex which is essentially a thioester. Thioesters make good substrates for nucleophilic attack because the sulfur is a good leaving group. Certain amino acids, including glycine and glutamine, can attack the acyl-CoA complex with their amino groups to form carboxyamides, which are amino acid conjugates of the carboxylic acids.

excellent nucleophile, coenzyme A (CoA), which contains a terminal thiol group. The resulting product is an acyl–CoA complex, but chemically it is a thioester. Thioesters react readily with nucleophiles with the elimination of a thioalkoxide. These make good leaving groups because sulfur can stabilize a negative charge due to its high polarizability (Chapter 1). The amino groups of the amino acids are the nucleophiles in this reaction and the process is catalyzed by an N-acyltransferase. One compound that undergoes this phase II process is the NSAID salicylic acid (Figure 7.51). The carboxylic acid group

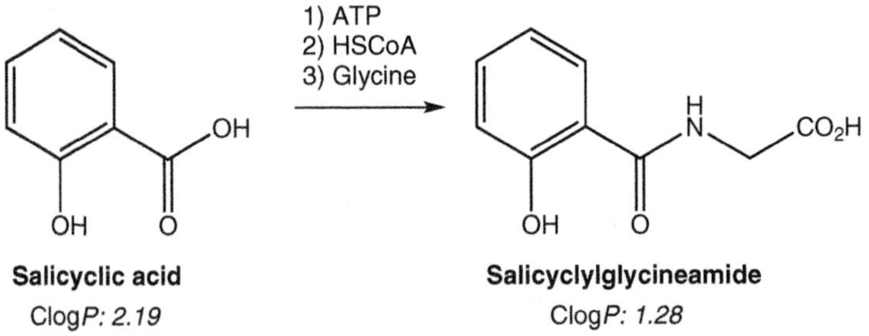

Figure 7.51 Salicylic acid is converted into a glycine conjugate mediated by ATP and coenzyme A. The resulting conjugate has a substantially lower ClogP value than salicylic acid. (ClogP values were determined using ChemBioDraw Ultra 13.0, Perkin Elmer Informatics, Waltham, MA).

is conjugated with glycine in the presence of N-acyltransferase to form the conjugate. The resulting carboxyamide is more water soluble than salicylic acid as determined by ClogP (1.28 vs 2.19 for the parent compound).

GSH conjugates are usually formed as a means of detoxification of reactive electrophilic species. Most often these electrophiles are formed during phase I metabolism by oxidative processes. Metabolites such as epoxides and arene oxides, as discussed earlier in this chapter, are highly reactive but are usually rendered harmless by reaction with water catalyzed by epoxide hydrolase, to form *trans*-dihydrodiols. Occasionally, however, such epoxides may have sufficient half-life so that other nucleophiles, such as functional groups associated with proteins, enzymes, or nucleic acids, are able to react before water resulting in covalently modified biomolecules with altered functionality. GSH is produced in the body to counter such threats. This is a tripeptide formed from glutamic acid, glycine, and cysteine, with the thiol group of cysteine being the most nucleophilic site.

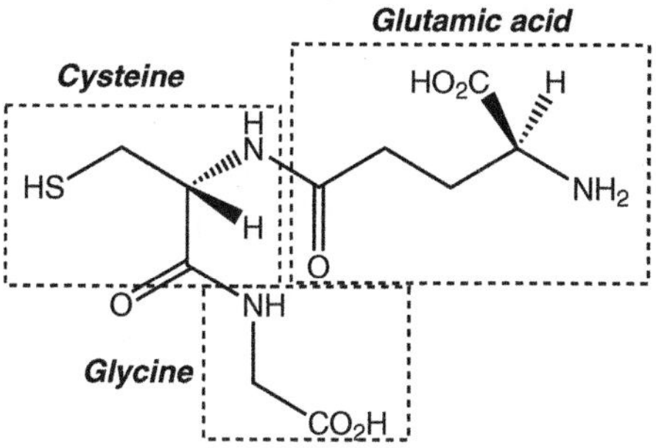

Glutathione

When an electrophilic metabolite is formed GSH is produced in sufficient concentration so that reaction with it becomes more likely than reaction with an essential nucleophilic biomolecule (Figure 7.52). The reaction between GSH and the electrophile is catalyzed by the enzyme glutathione S-transferase (GST).

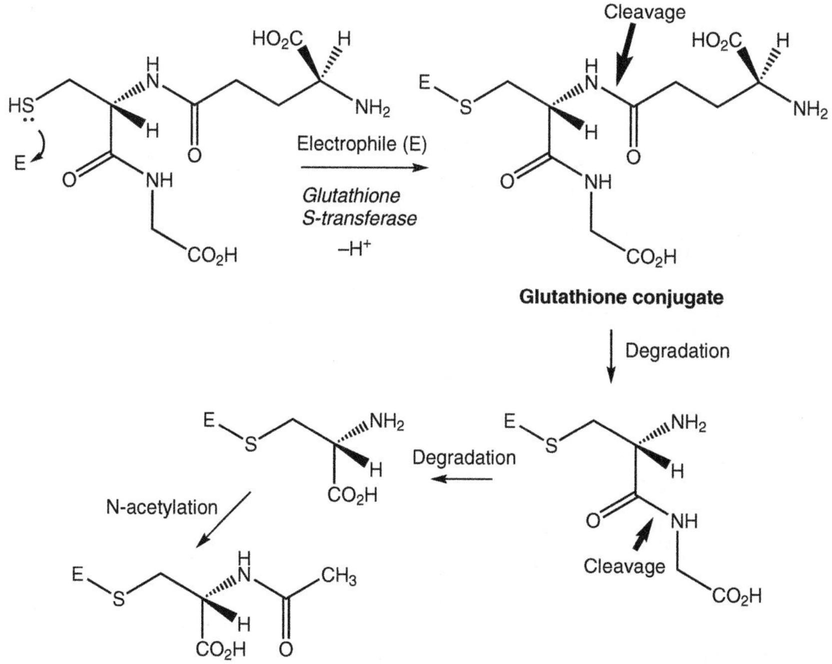

Figure 7.52 The thiol group of GSH reacts with electrophilic metabolites in the presence of GST to form a conjugate. These usually undergo degradation by cleavage of the glutamic acid and then the glycine residues. The resulting S-modified cysteine is usually acetylated on the amine to give the final, deactivated conjugate.

Once the reaction with GSH has occurred the toxicity of the electrophile is usually significantly reduced. The GSH conjugate then undergoes enzymatic degradation by cleaving the glutamic acid and glycine moieties. As a final step the amino group of the cysteine residue is often acetylated (see next section).

The reaction between GSH and electrophiles can occur by two main mechanisms—nucleophilic substitution or nucleophilic addition to an electron-deficient double bond. Nucleophilic substitution at an arene oxide, for example, requires that GST form a complex with a lone pair of electrons on the epoxide oxygen. The developing positive charge weakens one of the C–O bonds. GSH then approaches from the opposite side as the enzyme complex for steric reasons. Electrons from the sulfur form a bond to carbon with simultaneous cleavage of one C–O bond. Thus the result is a compound having a SGH⁺ group attached to one carbon of the former arene oxide with a *trans*-O-GST group on the other carbon. As the enzyme dissociates, a proton is transferred from the sulfur to oxygen to complete the reaction. This is illustrated for benzo[a]pyrene-4,5-oxide in Figure 7.53.

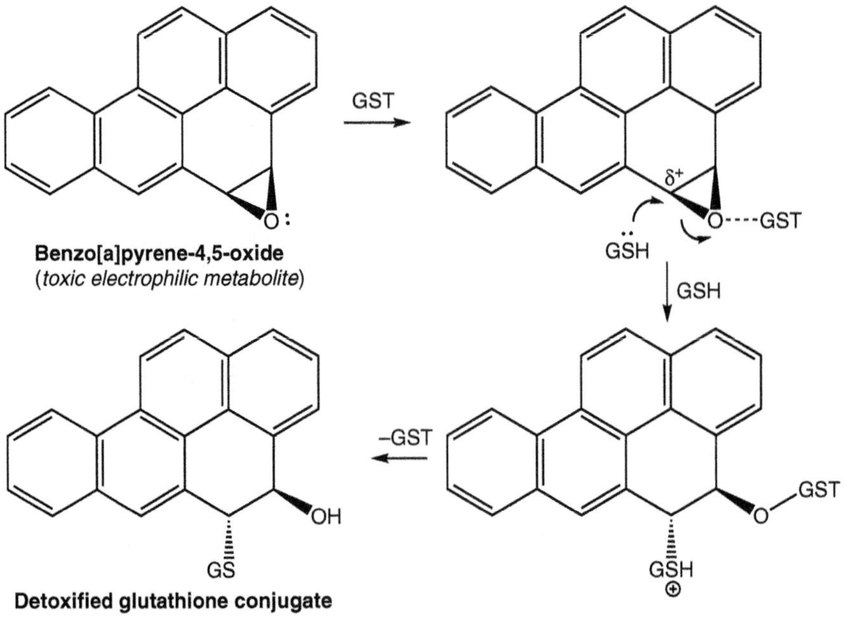

Benzo[a]pyrene-4,5-oxide
(*toxic electrophilic metabolite*)

Detoxified glutathione conjugate

Figure 7.53 Benzo[a]pyrene undergoes aromatic oxidation to produce, among other metabolites, a 4,5-arene oxide. GST coordinates with the epoxide oxygen which weakens the C–O bonds and makes it easier for GSH to attack from the opposite face to open the ring. Transfer of a proton from GSH to the oxygen completes formation of the detoxified conjugate. This is an example of nucleophilic substitution by GSH.

Acetaminophen (Tylenol®) is an over-the-counter drug widely used in children and adults for relief of pain and fever. It is also found as one component of many cold remedies. Individuals with a cold or the flu often take Tylenol to relieve their aches and pains and to lower fever and then might take a cold remedy to help them sleep at night, unaware that in so doing they are exceeding the recommended dosage of acetaminophen. This seemingly innocuous drug, however, undergoes a unique oxidative metabolism due to the relative orientation of the amide and phenol groups (Figure 7.54). The fact that they are *para*

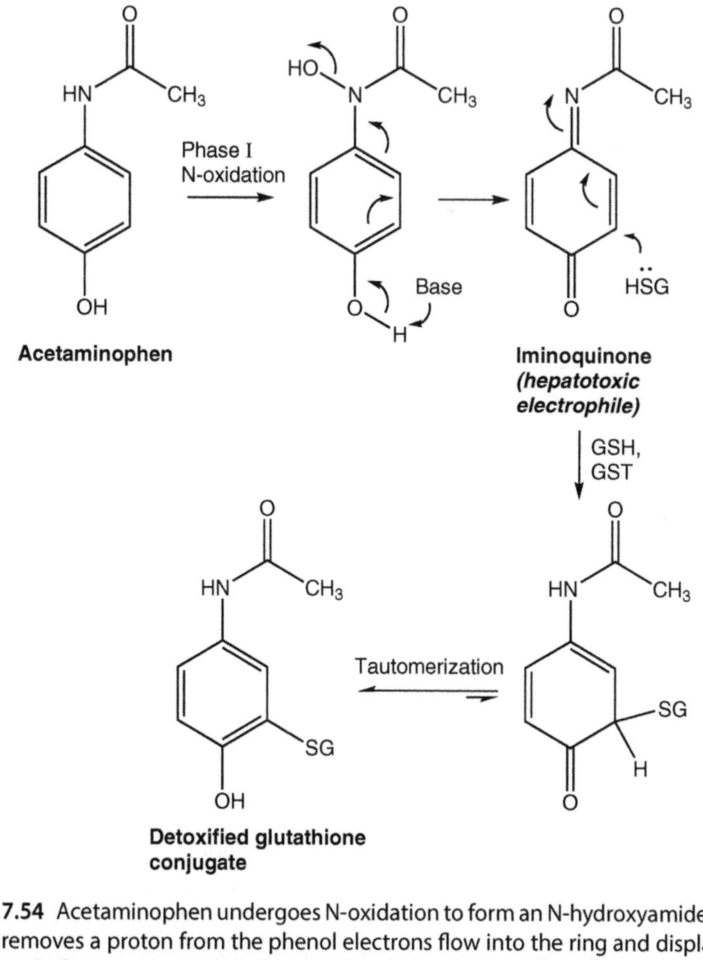

Figure 7.54 Acetaminophen undergoes N-oxidation to form an N-hydroxyamide. When a base removes a proton from the phenol electrons flow into the ring and displace the OH group to form an electrophile known as an iminoquinone. This is known to be toxic to the liver when formed in large quantities, such as after an overdose of acetaminophen. Normally, however, there is sufficient GSH present to detoxify this metabolite when the drug is taken in the prescribed amount. In this case GSH adds to the electrophile, aided by GST, to give an intermediate that tautomerizes to a phenol, but with GSH covalently bound to the 2-position.

allows for oxidation to a species known as an *iminoquinone*, formed by dehydrogenation of the NH and OH groups (this would not be possible if the two functional groups were in a *meta* arrangement). This may occur by oxidation of the amide nitrogen to an N–OH species as described earlier in the chapter. Removal of a proton from the acidic phenol group then pushes electrons into the ring, forming the carbonyl group. The excess electron density in the ring then forces elimination of the OH (or OR group) from nitrogen to complete formation of the iminoquinone. This species is no longer aromatic and has two double bonds that are in conjugation with both a carbonyl and an imine group. It is an electrophilic metabolite that if allowed to linger is toxic to the liver (hepatotoxic). The position adjacent to the carbonyl is most electrophilic because the imine nitrogen is further substituted by an electron-withdrawing acetyl group. Thus nucleophiles attack at this position and displace a pair of electrons within the ring forcing electron density onto the nitrogen. The resulting dienone can then tautomerize by shifting a hydrogen from the nucleophile-bearing carbon onto the oxygen reforming a phenol and reestablishing aromaticity in the ring. However, the nucleophile is now covalently bound to the 2-position of the ring. When taken at the proper dosage this iminoquinone metabolite does little damage, thanks to GSH. There is sufficient GSH present to handle the normal amount of iminoquinone metabolite and render it harmless. If an overdose of acetaminophen is taken, however, the amount of iminoquinone that is formed can exceed the amount of GSH available to neutralize it and liver damage can occur.

MISCELLANEOUS PHASE II PROCESSES

This final section of phase II metabolic pathways covers reactions that do not append large polar groups. Primary amines, whether present in the parent compound or produced via reductive metabolism, tend to be rather reactive since they are nucleophilic. One phase II process that primary amines tend to undergo is acetylation to form amides. In amides the nucleophilicity of the nitrogen is severely curtailed and thus acetylation serves as a means for terminating the activity of and detoxifying primary amines. Unlike other phase II conjugative processes the resulting amides are not necessarily more water-soluble than the amines from which they derive. The reaction is analogous to that used to form glycine and glutamine conjugates of carboxylic acids, only here the acid is acetic acid and it is an ordinary amine and not an amino acid that serves as the nucleophile (Figure 7.55). Thus the actual acylating agent is acetyl-CoA, and the reaction is catalyzed by an *N*-acyltransferase. Acetylation can occur on primary aliphatic and aromatic amines as well as on the NH_2 groups of sulfonamides and hydrazides. Several examples are shown in Figure 7.56.

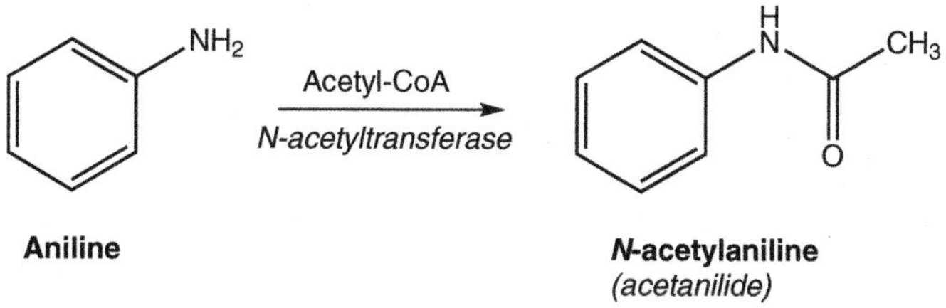

Aniline

N-acetylaniline
(acetanilide)

Figure 7.55 Primary amines often undergo phase II metabolism by appending an acetyl group onto the nitrogen to form an amide. This greatly diminishes the nucleophilicity of the nitrogen often preventing toxic effects that might have arisen from the amine. The reaction employs N-acetyltransferase to catalyze the reaction, which requires acetyl-CoA as a source of the acetyl group. Here the process is shown for an aromatic amine.

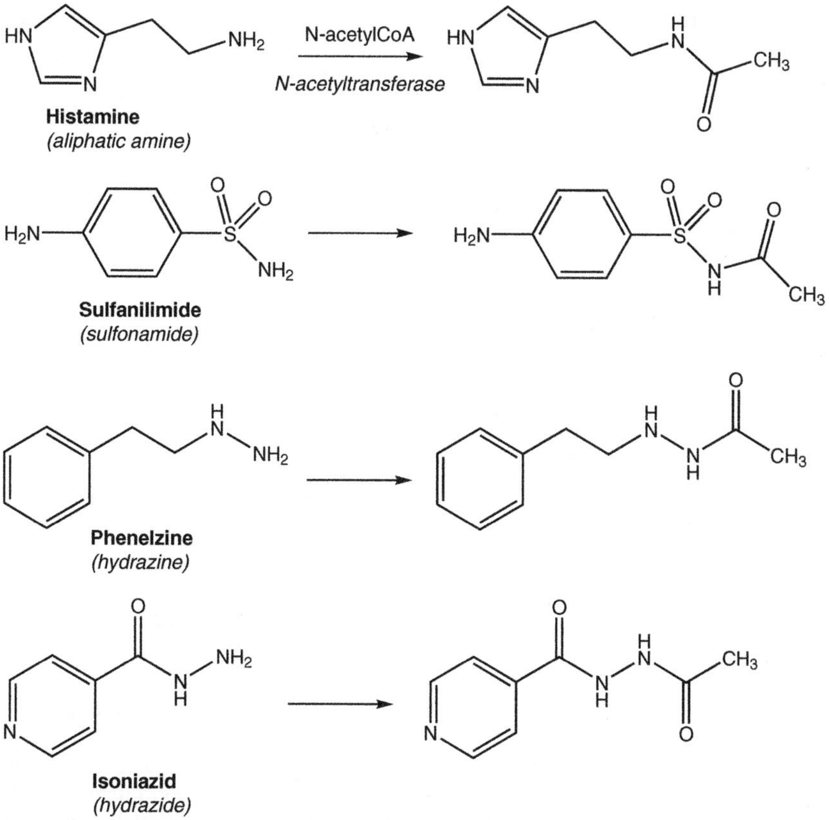

Histamine
(aliphatic amine)

Sulfanilimide
(sulfonamide)

Phenelzine
(hydrazine)

Isoniazid
(hydrazide)

Figure 7.56 Other NH$_2$ groups can act as substrates for N-acetylation. Shown here are several examples including an aliphatic primary amine, a sulfonamide, a hydrazine, and a hydrazide.

The final phase II metabolism to be discussed will be methylation. This is a relatively minor pathway that may occur with phenols, *catechols* (1,2-dihydroxyphenyl groups), amines, and thiols. As with acetylation of amines, methylation does not result in a more water-soluble metabolite, but it does generally terminate the activity of the functional groups to which it becomes attached. The compound *S*-adenosyl methionine (SAM) acts as the agent that transfers a methyl group and the reaction is catalyzed by various methyltransferases (Figure 7.57). Examination of the structure of SAM reveals a trisubstituted sulfur (sulfonium ion), which necessarily must be positively charged. A nucleophilic functional group uses a lone pair of electrons to attack the methyl substituent on the sulfonium ion. Simultaneous with this the bond connecting the methyl group to sulfur breaks, pushing the electrons onto the sulfur to reestablish a lone pair. Loss of a proton completes the formation of the methylated functional group.

Monomethylation of catechols is a common metabolic process, whereas methylation is only a minor pathway for ordinary phenols. For phenols the most common phase II metabolism is glucuronide formation. An example

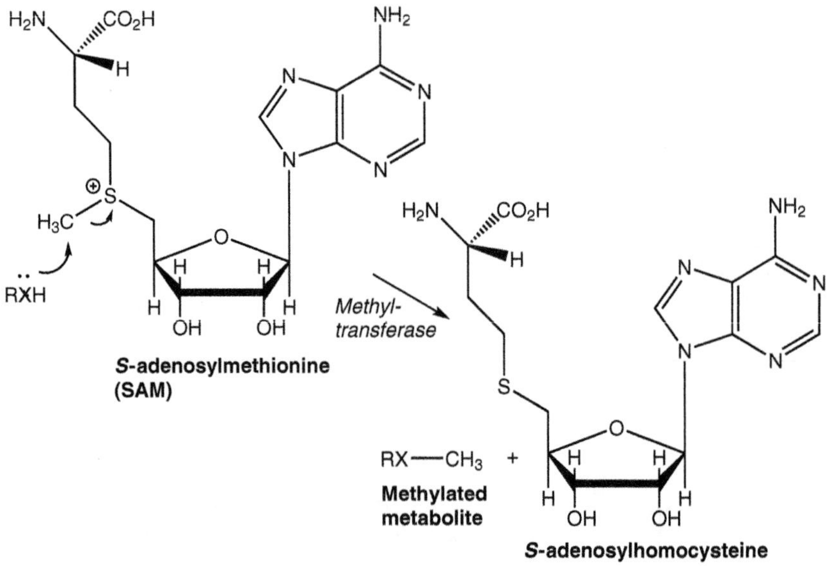

Figure 7.57 The compound SAM has a methyl sulfonium group (a trisubstituted sulfur with a positive charge). Nucleophilic functional groups such as phenols (usually catechols or 1,2-dihydroxyaryl compounds) and occasionally amines and thiols can attack the methyl group, catalyzed by methyl transferase, to give a methylated phase II metabolite after loss of a proton along with release of *S*-adenosylhomocysteine.

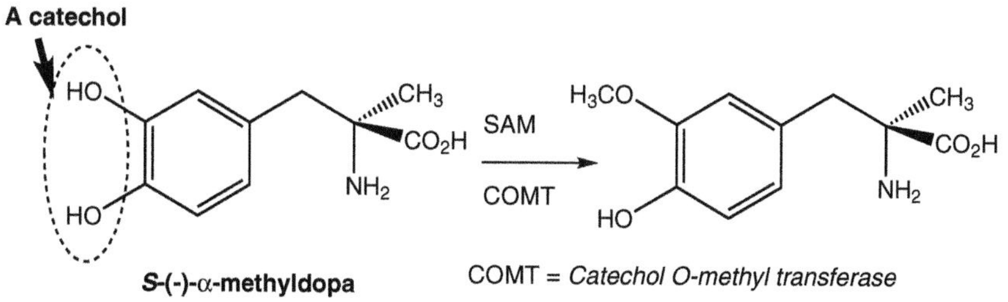

A catechol

S-(-)-α-methyldopa   COMT = *Catechol O-methyl transferase*

Figure 7.58 Catechols are the most common substrate for SAM. In this example the drug *S*-(-)-α-methyldopa is methylated by SAM to the 3-methoxy metabolite. The methyl transferase that catalyzes this process is called catechol *O*-methyl transferase.

of a catechol–containing drug is *S*-(-)-α-methyldopa (Figure 7.58). Monomethylation of the catechol is catalyzed by the enzyme catechol–*O*-methyl transferase (COMT).

REFERENCES

[1] G. Guroff, J.W. Daly, D.M. Jerina, J. Reson, B. Witkop, S. Udenfriend, Hydroxylation-induced migration: the NIH shift. Recent experiments reveal an unexpected and general result of enzymatic hydroxylation of aromatic compounds, Science 157 September 29 (1967) 1524–1530.

[2] T.L. Perry, C.F.A. Culling, K. Berry, S. Hansen, 7-Hydroxychlorpromazine; potential toxic drug metabolite in psychiatric patients, Science 146 October 2 (1964) 81–83.

[3] B. Entsch, D. Ballov, V. Massey, Flavin-oxygen derivatives involved in hydroxylation by *p*-hydroxybenzoate hydroxylase, Journal of Biological Chemistry 251 (9) (1976) 2550–2563.

[4] J.H. Weisburger, E.K. Weisburger, Biochemical formation and pharmacological, toxicological, and pathological properties of hydroxylamines and hydroxamic acids, Pharmacological Reviews 25 (1) (1973) 1–66.

[5] V. Massey, F. Mueller, R. Feldberg, M. Schuman, P.A. Sullivan, L.G. Howell, S.G. Mayhew, R.G. Matthews, G.P. Foust, Reactivity of flavoproteins with sulfite. Possible relevance to the problem of oxygen reactivity, Journal of Biological Chemistry 244 (15) (1969) 3999–4006.

[6] R.P. Mason, J.L. Holtzman, Mechanism of microsomal and mitochondrial nitroreductase. Electron spin resonance evidence for nitroaromatic free radical intermediates Biochemistry 14 (8) (1975) 1626–1632.

[7] R.P. Mason, F.J. Peterson, J.L. Holtzman, The formation of an azo anion free-radical metabolite during the microsomal azo reduction of sulfonazo III, Biochemical and Biophysical Research Communications 75 (3) (1977) 532–540.

INDEX

Note: Page Numbers followed by f indicate figures; t, tables; b, boxes.

Printed and bound by CPI Group (UK) Ltd, Croydon, CR0 4YY

03/10/2024

01040421-0001